# Warship 2021

AA005728

## Future Technologies in Naval Submarines

Online
2 - 3 June 2021

ISBN: 978-1-7138-8308-1

**Printed from e-media with permission by:**

Curran Associates, Inc.
57 Morehouse Lane
Red Hook, NY 12571

**Some format issues inherent in the e-media version may also appear in this print version.**



The Royal Institution of Naval Architects
8-9 Northumberland Street
London, WC2N 5DA
United Kingdom

Phone: 020 7235 4622
Fax: 020 7259 5912

publications@rina.org.uk

**Additional copies of this publication are available from:**

Curran Associates, Inc.
57 Morehouse Lane
Red Hook, NY 12571 USA
Phone: 845-758-0400
Fax: 845-758-2633
Email: curran@proceedings.com
Web: www.proceedings.com

**International Conference**

# Warship 2021 Future Technologies in Naval Submarines

2-3 June 2021

**PROCEEDINGS**

THE ROYAL INSTITUTION OF NAVAL ARCHITECTS
8-9 Northumberland Street, London, WC2N 5DA, Telephone: +44 (0)20 7235 4622

The Royal Institution of
Naval Architects

INTERNATIONAL CONFERENCE

# Warship 2021 Future Technologies in Naval Submarines

2-3 June 2021, Online Conference



Papers published in these conference Proceedings are peer reviewed.

THE ROYAL INSTITUTION OF NAVAL
ARCHITECTS
8-9 Northumberland Street
London WC2N 5DA

Telephone: 020 7235 4622
Fax: 020 7259 5912

ISSN No: 2752-6569

# CONTENTS

# UNLOCKING THE POTENTIAL OF SUSTAINABLE SUBMARINE DESIGN

**L Hammock,** BMT, UK


## SUMMARY

Submarines are designed for a service life of more than 25 years and therefore they need to be designed to meet the needs of tomorrow as well as today. This not only means a flexible operational capability but also implies increased importance should be placed on areas such as sustainability and climate resilience in design. Although submarines, through low signature requirements, often present a lower environmental footprint than their larger ship counterparts there is still a considerable concerted effort required to ensure the platform is environmentally sustainable, throughout its entire life: from concept to end of life disposal. This paper will explore the key drivers and factors behind environmental sustainability in submarine design, with examples demonstrating how sustainable principles may be used to improve the environmental impact of the platform, without compromising the mission and whilst improving performance. The paper will also discuss the role of teamwork in achieving this goal, including how designers, customers, shipyards and suppliers can all work together to maximise the benefits of a sustainable submarine design.


## 1. INTRODUCTION

The term 'sustainability' is increasingly used but can have differing interpretations. The United Nations (UN) Bruntland Commission Report, (UN, 1987), defines sustainable development as "development that meets the needs of the present without compromising the ability of future generations to meet their own needs.".

The UN has identified seventeen Sustainable Development Goals (SDGs) (UN, 2021): recognising the need to work together globally and holistically. Many of the SDGs require a change of mindset and behaviour, for successful implementation.

*Figure 1 – UN Sustainable Development Goals - Figure reproduced from the UN (2021)*

Focusing in on Environmental Sustainability; it is more than simply Environmental Protection. There are various technologies available to reduce a platform's immediate impact upon the environment, however, to be Environmentally Sustainable the designer must assess the environmental impact over the whole lifecycle.

## 2. SUSTAINABILITY IN DEFENCE

An important consideration for Defence activities and assets is the ability to remain operationally effective. But this can be achieved in synergy with environmental sustainability. In fact, following a sustainable approach to asset design, procurement, operation, and disposal, can also provide opportunities for improvements and more efficient operation throughout the platform lifecycle. This was recognised in the recently published Ministry of Defence Climate Change and Sustainability Strategic Approach (MoD, 2021); "In widening and embedding sustainability, we will look at our impact on the environment in a whole different way. This will reveal opportunities for building greater resilience and creating new types of partnership."

Climate change has, and will continue to have, an impact on the way that Defence, as a whole, protects, operates and fights. All sectors of Defence have a part to play in reducing the contribution to climate change and biodiversity loss.

## 3. ASSET LIFECYCLE

The asset lifecycle is considered to follow the path of Concept, Assessment, Demonstration, Manufacture, In-Service, Disposal (CADMID). Using this lifecycle framework allows the designer to assess and minimise the overall impact upon the environment. It applies equally to components, equipment, systems, and platforms.

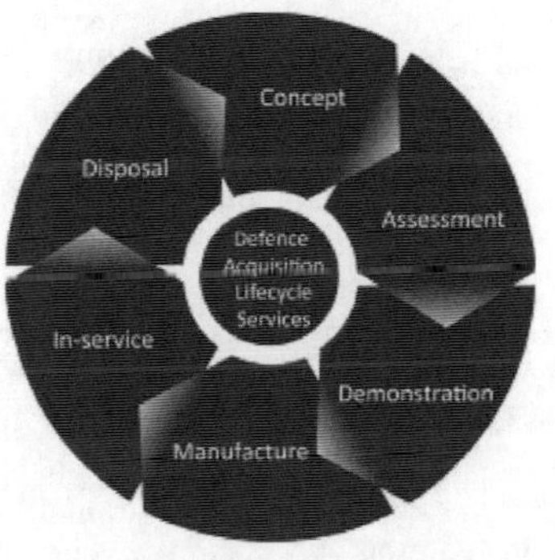


*Figure 2 – CADMID lifecycle reproduced from Interaction (2021)*

By undertaking an Environmental Impact Assessment (EIA) early in the lifecycle, the full environmental impact of the platform can be understood and mitigated. The impacts of the production, transportation and disposal aspects of a programme are often overlooked yet are easily designed-out during the early phases and by reassessing throughout the lifecycle, rather than attempting to remedy the impacts of equipment or platforms once in service. By starting with the end in mind, the design of the platform may be optimised throughout its life.

An EIA should be conducted at key milestones throughout the design process (for example, at design reviews and when a key design change is made), to assess the impact of the platform at each phase of its lifecycle. The cost of designing in mitigations at the front-end is almost always lower, and simpler, than attempting to rectify later. Refrigerant systems are a good example of this. The 2014 EU F-Gas Regulation (517/2014) controls the use and emissions of F-Gases. Hydrofluorocarbons (HFCs) will be phased down in the UK, to achieve a reduction of 79% by 2030, compared with average use between 2009 – 2012. Systems and equipment suitable for lower Global Warming Potential (GWP) refrigerants can be significantly different from those used in previous designs and can be complex and expensive to retrofit. Given the length of the CADMID cycle of a platform, consideration should be given to the likely availability of refrigerants during the platform's life, in order to design out issues of obsolescence, or increased cost, at a later stage. It may be reasonably assumed that some of the common refrigerants being used today with higher GWP will be phased out during the lifecycle of a platform. This can lead to increased operating costs during the phase out period, before the refrigerant becomes unavailable. The financial and environmental costs of retrofitting a different system during the platform's life must be considered at the front-end of the CADMID cycle, to allow the designer to investigate alternative refrigerants and systems, to future-proof the design.

Many designers are now considering ammonia and CO2 as viable alternatives to higher GWP refrigerants. Ammonia and CO2 have inherent safety hazards, however these hazards may be mitigated early in the platform design phase, making these options feasible. When selecting these options, it is important to investigate the supply chain of the ammonia or CO2: historically, the production of ammonia involved high temperatures and pressures, and was a very energy intensive process. The traditional Haber-Bosch method of producing ammonia is estimated to contribute around 1% of global Greenhouse Gas (GHG) emissions (Brown, 2016). However, greener methods of ammonia production are in development, which could improve its overall carbon footprint and make it a more sustainable product, as well as having a lower direct environmental impact.

Similar supply chain analysis should be undertaken when considering any system design or design change: including future fuel selections. Working closely with suppliers and shipyards to reduce the environmental impact throughout the supply and production chains, the further reaching environmental impact of the platform may be reduced.

## 4. WATER TREATMENT STREAMS

There are elements that the designer might consider when designing systems, to enable the end user to operate the platform in a more environmentally sustainable manner, during the in-service part of the platform lifecycle. Bilge management is one area where good operating practices may be encouraged through the design of the system. There are two types of bilge water that accumulates on ships and submarines: clean and oily.

Discharge of oily water is regulated under MARPOL Annex I (IMO, 1983). Bilge water with an oil content greater than 15 parts per million (ppm) shall not be discharged overboard. From a mission perspective, a submarine could also be compromised by oil sheen on the water surface, by discharge overboard of bilge water with oil content. Therefore, effective bilge management is important on a submarine from the perspective of environmental sustainability and operational sustainability.

Designing for a dry bilge concept, for example, involves the piping designer routes clean drainages directly into small collecting tanks in the bilge. Designing this in from the start means that the contents of these small tanks is guaranteed to be oil free and so there is no risk of compromise to the environment or the mission when discharged.

Because the remaining bilge water with potential oil content will be a smaller volume, the designer will have increased options available for treatment or storage solutions.

## 5. WASTE NOT WANT NOT

Another example of using the design process to instil sustainable practises during the in-service phase, is reducing wasted energy.

When designing exothermic processes, designers should consider the possibilities of harnessing waste heat. Waste heat recovery systems do not have to be large and, when designed in from an early stage, may be easily integrated into a compact system design. Waste heat may be used for various applications such as heating water or spaces. This careful use of energy is particularly applicable to diesel powered submarines. In addition to energy recovery, a mindset to reduce energy consumption on board is also important. The responsibility for sustainable submarine operations belongs to everyone on board, as

well as the design and procurement teams. Switching off lights and non-essential equipment will help in the overall energy savings. Energy consumption reduction strategies in operation are also beneficial for the mission, as well as the environment: operational sustainability and environmental sustainability often work together.
For the designer: low energy equipment may be designed in. Low energy lighting, reduced flow shower heads, low flush toilets and treated grey water reuse on board for non-potable purposes are all ways to design in a manner that encourages environmentally sustainable operations in-service. By designing sustainably from the front-end, improvements to the maintainability of equipment and therefore operations are often also improved.

With regards to nuclear powered submarines; considering the whole lifecycle provides a wider understanding of the 'hidden' environmental impact of nuclear power: the mining and processing of uranium and the radioactive waste handling should be considered when assessing the overall environmental impact of the platform. Whilst nuclear energy is often considered to be 'clean' and 'free' in environmental terms, it should still be used mindfully. When using energy on board a submarine, the overall environmental 'cost' per kW of energy should be kept in mind.

## 6. STARTING WITH THE END IN MIND

An important, yet often overlooked, part of the lifecycle is disposal. Disposal impacts are evident throughout the life of the platform. There are operational wastes generated from product or equipment packaging and waste generated by personnel. Waste is produced during design changes, when older equipment is removed, and in the packaging of new equipment to replace it. Finally, waste is generated at the end of the platform lifecycle, when the platform and its assets are removed from service.

Operational waste can be significantly reduced through careful analysis of the supply chain and working closely with suppliers. Increasing numbers of production companies are now looking at their impacts on the environment and sustainable production. Many are finding ways to reduce the volumes of packaging used and to use recyclable packaging. Selecting suppliers with sustainable credentials, and working together with suppliers and manufacturers to maintain and improve these credentials, will reduce the waste associated with the platform.

Operational waste should always be separated to allow recycling when disposed of ashore. But the responsibility does not end there: the waste contractor should also be carefully selected for their 'green' credentials, to ensure that the waste generated by the platform will be handled in a sustainable manner.

There are some inherently difficult wastes generated on board a platform, which are difficult to store safely during a mission. Food waste and, where women are present on board, sanitary waste present a health hazard if stored incorrectly. When considered from the initial design phase, there are several solutions to a reduced waste production: from chilled storage, to dehydrators and auto-gasification units, which produce an inert, compact by-product for standard storage and landing ashore. More recent waste handling technologies are compact and quiet: making them ideal alternatives for submarines.

Submarine endurance is a key driver of operational effectiveness, and waste storage may be a limiting factor. By considering the whole life-cycle of waste on the platform: i.e. reducing the packaging of consumables to allow more stores, processing the waste in a way to safely minimise it's stored volume onboard; it may be possible to extend patrol length.

By assessing the volumes and types of waste anticipated, at an early design phase, and through the careful selection of the most appropriate equipment for the type and volume of waste, sufficient space may be allocated from the outset to provide an appropriate solution.

Ultimate disposal of the platform and onboard equipment should be considered throughout the entire CADMID cycle of the platform. Whether equipment is being procured in the design phase or as a design change part way through the life of the platform, consideration should be given to how the materials will be handled at the platform or equipment end of life.

The Sustainable MOD Strategy Waste Management 2015 – 2025 (MoD, 2015) follows the waste hierarchy set out in the European Waste Framework Directive, as transposed into UK law by The Waste (Circular Economy) (Amendment) Regulations 2020 (UK Gov, 2020). This hierarchy declares the highest priority is the prevention of waste, followed by reuse and recycle. These elements should be explored with prospective suppliers to ensure that sustainable disposal is planned for at an early stage.

As well as designing for reduced waste and reusable or recyclable materials, the impact of waste should also be considered. Whilst toxic materials are inherently restricted onboard a submarine, some materials are hazardous to the environment, or require energy intensive processes to recycle or dispose of them. By avoiding the use of such materials, the overall environmental footprint of the platform is improved. Where the use of these materials cannot be avoided they must be documented, along with volumes and locations, and a plan put in place for handling their disposal.

The security aspects of components and equipment should also be considered at the design phase to ensure that any security restrictions upon recycling / reuse are included in the disposal plan. Creating a disposal plan at an early point in the design phase, and maintaining the plan throughout the lifecycle, will ensure a clear picture of waste management and reduce the amount of waste consigned to disposal processes.

This is increasingly important for nuclear powered submarines, which require a coherent strategy for managing laid-up submarines. In particular, the handling and storage of High Level Nuclear Waste.

## 7. CONCLUSIONS

Environmental Sustainability involves a holistic view of the platform across its lifecycle. It is about a mindset shift in the design, operation and disposal of a submarine. There are many systems and equipment available to reduce the environmental impact of a platform. However, often these options are not practicable for installation on board a submarine, due to space constraints, noise restrictions, weight considerations and atmospheric conditions. By using sustainability principles from an early stage in the design process, elements of the platform requiring environmental impact mitigation can be reduced by design, resulting in a greater selection of environmental protection equipment, more suitable to the submarine environment.

Applying sustainability principles further, across the entire lifecycle of the platform, it is possible to reduce the environmental impact of a submarine, without compromising its mission capability. Indeed, in many of the examples cited in this paper, implementing sustainability practices at the front-end of the design process is also beneficial to operational effectiveness.

## 8. ACKNOWLEDGEMENTS

The authors would like to acknowledge all those who supported the research conducted at BMT and the production of this paper.

## 5. REFERENCES

1. Brown, T., 2016. The Investment Case for Sustainable Ammonia Synthesis Technologies. In: 13th Annual NH3 Fuel Conference.

2. Eur-lex. 2021. EUR-Lex - 32014R0517 - EN - EUR-Lex. [online] Available at: <https://eur-lex.europa.eu/legal-content/EN/TXT/?uri=uriserv%3AOJ.L_.2014.150.01.0195.01.ENG> [Accessed 6 May 2021].

3. IMO, 1983. MARPOL Annex I – Prevention of Pollution by Oil. [online] Available at: <https://www.imo.org/en/OurWork/Environment/Pages/OilPollution-Default.aspx> [Accessed 6 May 2021].

4. Interactionofbath.com. 2021. Interactionofbath. [online] Available at: <http://www.interactionofbath.com/> [Accessed 6 May 2021].

5. MoD, 2015. Sustainable MOD Strategy Waste Management.

6. MoD, 2021. Climate Change and Sustainability Strategic Approach. [online] MoD. Available at: <https://assets.publishing.service.gov.uk/government/uploads/system/uploads/attachment_data/file/973707/20210326_Climate_Change_Sust_Strategy_v1.pdf> [Accessed 6 May 2021].

7. The Waste (Circular Economy) (Amendment) Regulations 2020. 2020 No. 904.

8. UN, 1987. Report of the World Commission on Environment and Development :. [online] United Nations Digital Library System. Available at: <https://digitallibrary.un.org/record/139811> [Accessed 6 May 2021].

9. UN, 2021. THE 17 GOALS | Sustainable Development. [online] Sdgs.un.org. Available at: <https://sdgs.un.org/goals> [Accessed 6 May 2021].

# APPLYING MODERN PROCESS AUTOMATION AND OPTIMISATION TOOLS TO THE DESIGN OF SUBMARINE STRUCTURES

**D Graham, A White, S Cross, C Donaldson, M Lytwyn** and **R Craven,** QinetiQ Rosyth, UK

## SUMMARY

This paper describes the application of commercially available process automation and optimisation software to the design of submarine pressure hulls. The *Isight* software provided by Dassault Systemes, which integrates with a range of internal and external tools, including their *Abaqus* FE software, and provides access to numerous sophisticated design assessment and optimisation routines, was used. Within the *Isight* environment, disparate tools can be linked in structured workflows and the following case studies are described:

- Minimum weight pressure hull design for different operating depths using traditional design formulae.
- Design refinement for weight reduction using linear elastic FE.
- Sensitivity of pressure hull collapse strength to shape imperfection using elasto-plastic non-linear FE.

Automating these processes does lead to an increase in the amount of output, which can require an experienced eye to interpret, but they have produced demonstrable design improvements that just would not have been realised by manual methods.

## 1. INTRODUCTION

The current authors' direct predecessors were heavily involved in the development of the design rules for submarine pressure hulls when the Rosyth site was one of the MoD's research establishments. In fact the first full summary of the UK knowledge of the design of submarine hulls for static strength was written in an internal report of the Naval Construction Research Establishment (NCRE) by Bill Kendrick shortly before the oldest of the current authors was born. A modified version of that report was later published openly in the book edited by Gill in 1970 [1]. Kendrick later added an elasto-plastic overall collapse analysis and many others have contributed to development since, but the basic design philosophy remains unchanged.

There are a number of reasons for the longevity of this process: it is based on a rigorous analytical background backed up by extensive testing and over many years it has been used to produce efficient, reliable designs for pressure hulls of UK submarines. As part of an ongoing initiative MoD and QinetiQ have been working together to gather legacy design tools together in a more modern, maintained and supported Windows based environment. This has resulted in the development of the Submarine Structural Tools (SST), which is currently being tested in industry. Part of this process involved a thorough review of current design standards and legacy documentation. This was important as a lot of deep knowledge has been, if not lost, at least partially obscured by the passage of time.

If the development of the SST is making legacy design tools more accessible to a new generation of designers, the scope of design is still largely restricted to the philosophy established over fifty years ago. The success of this approach has also, perhaps, restricted the adoption of advanced numerical techniques such as Finite Element (FE) analysis in overall design, although work over the last ten or fifteen years [2], [3] & [4], for example, has gradually increased the confidence in FE analysis and helped justify its use in a range of assessment applications.

Formal optimisation is not new and has been investigated in the past but it could require the integration of multiple tools from different suppliers and although a lot of good work was done, it was often not easily portable to general applications. However, improved availability of more integrated optimisation software and the increased confidence in FE analysis described above has prompted a revisit of structural optimisation and design analysis.

In this paper the *Isight* software [5] has been used to drive studies aimed at optimising the design of pressure hulls, using elements of the SST design tools and a mix of linear and non-linear FE analyses.

## 2. ISIGHT TOOLS AND WORKFLOWS

*Isight* is a process integration and automation tool for managing simulation based design analysis. It allows the user to set up and manage various software components, ranging from simple scripts and spreadsheets to non-linear FE, and to apply advanced techniques such as formal optimisation, Design of Experiments (DOE) and Design for Six Sigma (DFSS). It is based around a visual environment where the user can drag and drop tools and components, define connections and set up process flows. These can range from basic loops, as shown in Figure 1, to complex flows with multiple branches, nested sub-loops etc.

 

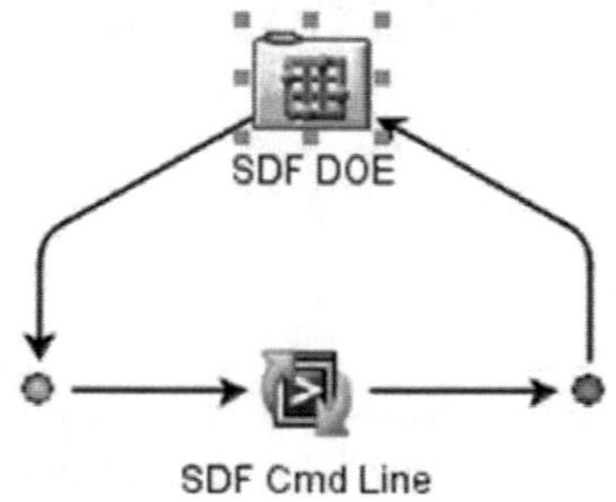


Figure 1. A simple *Isight* workflow loop.

## 2.1 COMPONENTS AND TOOLS

The menu of application components in *Isight* includes native tools and third party applications ranging from MS Word and Excel to Abaqus, SolidWorks and CATIA. This is entirely customisable and the user can add additional components if required.

The 'drag and drop' graphical workspace can belie a hidden complexity in the apparent simplicity of links between components. To accommodate the disparate software that can be linked, the protocols for data transfer are open ended, and give access to a wide range of variables and parameters. These are accessed in a sub-menu for a component.

For example, the 'Simcode' component shown in Figure 1 'SDF Cmd Line' has been set up to extract all the relevant inputs from a text file, run an executable which carries out a Submarine Design Formulae (SDF) analysis and return the relevant outputs. It can be quite tricky to establish the required relationships and to verify that they are working correctly as intended. However, this flexibility gives the user a very powerful tool to develop his or her own applications.

*Isight* provides a menu of process components which drive the analysis. In the current work we have used Design of Experiment (DOE), as in figure 1, and Optimisation.

DOE is a method for exploring a design space where the user identifies a set of variables and sets a range and number of levels for each to be assessed at. The most accurate and reliable technique is 'Full Factorial' which executes a calculation for every combination but this can be expensive for multiple factors at multiple levels. For example 5 factors studied at 5 levels requires 3125 individual calculations. The DOE component in *Isight* provides methods for identifying a reduced data set which will provide an acceptable answer, such as the Optimal Latin Hypercube method.

*Isight* offers 17 separate optimisation techniques. Details of the optimisation techniques are given in the *Isight* documentation [5] but they cover a range of techniques including gradient based methods and genetic algorithms, each suited to the solution of different types of optimisation problem.

*Isight* also provides tools for monitoring an analysis as well as graphical displays and other post processing capabilities.

## 2.2 WORKFLOWS

To fully capitalise on the power of the tool, there is an obvious investment in the development of robust, proven workflows. As part of the work described in this paper three bespoke workflows were developed and tested, to a point where they could be reliably shared with other users. These are described below and independent case studies using each workflow are described in Section 3.

### 2.2 (a) Submarine Design Formulae

The Submarine Design Formulae (SDF) are a component of the SST described earlier. A command line interface to the calculation and post-processing part of the tool was developed and included in the workflow shown in Figure 1. This gives the user access to all of the inputs (dimensions, scantlings and material properties of a pressure hull compartment) and outputs (collapse pressures, elastic buckling pressures, limiting stresses etc.) of the documented validated SDF tool, which users have experience and confidence in using. It also allows users to manually re-run the same calculation interactively if required.

The SDF tool is normally used at the early design stage or for quick assessments of at-build or in-service conditions and, while the graphical SST is quick and easy to run, it can be difficult to keep track of multiple runs. The *Isight* workflow makes it particularly easy to set up sensitivity studies, potentially involving thousands of runs, and collate and present the results in a coherent fashion.

### 2.2 (b) Linear FE

The workflow shown in figure 2 was developed to provide a refined assessment or optimisation based on linear elastic FE analysis. The approach taken was to place design limits based on elastic buckling pressure and peak stresses. To access these outputs, it was necessary to carry out these analyses separately but this also meant that the buckling and stress analysis models could be tailored. For example a less refined mesh was adequate for the buckling analysis.

 

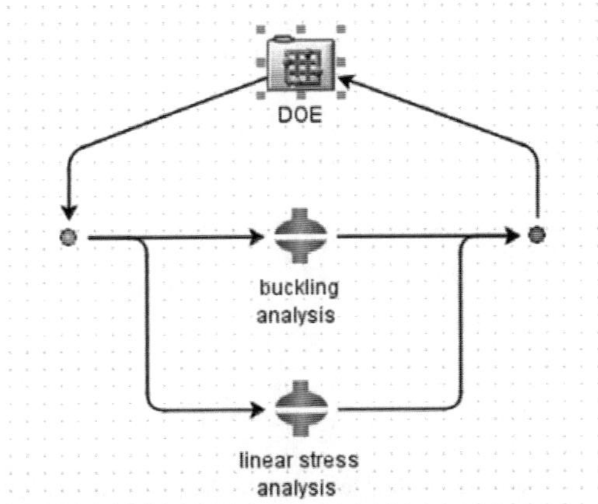


Figure 2. Workflow for elastic FE.

Accessing the FE model database directly gave access to fundamental dimensions, held in the model sketch, as well as shell thicknesses. An application of this workflow is discussed in Section 3.2.

2.2 (c) Non-Linear FE

The prediction of pressure hull collapse using FE requires non-linear analysis and is particularly sensitive to initial shape imperfections. QinetiQ have developed some simple to use tools to generate basic hull models and introduce generic shape imperfections based on idealised overall and interframe imperfections including those described by Graham et al. [4]. These are also graphical interface based tools but it was a relatively simple matter to create a command line interface to access the functionality within *Isight* and the basic workflow is the same as that shown in Figure 1.

2.3 SUMMARY

The workflows described in this section represent a significant effort but now provide a robust baseline for developing bespoke studies. They can be picked up by a new user and used with confidence that they will perform as expected (at least on the same computer hardware). The user is free to add additional components, change the controlling tools or generally adapt the workflow to a specific application; and to return to the original at any time. This is illustrated in the following case studies.

## 3. CASE STUDIES

### 3.1 OPTIMISED DESIGN OF 'SHALLOW' AND 'DEEP' DIVING HULLS USING SST.

For this exercise the DOE application shown in Figure 1 was replaced with formal optimisation. This process was used to generate preliminary designs for notional 'deep' and 'shallow' diving pressure hulls. Bounds were set on the interframe collapse pressure, overall collapse pressure, frame tripping pressure and the frame yield pressure. These were arranged to give target operating depths of approximately 2000 m and 200 m respectively.

The overall dimensions of a Notional Compartment were taken as a starting point and in each case some work was done to define suitable ranges for the design variables for each design study (for example larger thicknesses were required for the deep diving design). Hull thickness, frame spacing, and the frame scantlings were all allowed to vary.

Because most of the optimisation procedures could be run in a matter of minutes a full study of all the options was carried out. It was found that some performed better than others and that, although several converged on similar minimum weights, the final designs could be quite different. This is perhaps unsurprising given the complexity of the design space.

Design optimisations were carried out separately with three different materials Titanium, Steel and Aluminium to represent a mix of high strength / light-weight materials. The properties used are summarised in Table 1.

| Property | Titanium | Steel | Aluminium |
|---|---|---|---|
| E (GPa) | 114000 | 207000 | 70000 |
| $\sigma_y$ (MPa) | 1000 | 550 | 300 |
| $\rho$ (kg/m3) | 4400 | 7800 | 2700 |

Table 1. Material properties.

Minimum weight solutions from the more successful optimisation routines, for the deep diving designs, are given in Table 2.

| Method | Titanium | Steel | Aluminium |
|---|---|---|---|
| LSGRG | 616.2 | 1799 | 1183 |
| MISQ | 630.6 | 1784 | 1131 |
| NLPQLP | 625.8 | 1802 | 1145 |
| Multi-Island GA | 872.8 | 1913 | 1159 |
| Multi-objective PS | 906.3 | 1856 | 1179 |
| Pointer | 633.4 | 1784 | 1130 |

Table 2. Minimum weights (T) for deep diving designs.

The obvious conclusion is that titanium, with its high strength and relatively low density is the best material for a deep diving hull, at least from purely structural considerations. In theory there is nothing to prevent other factors, such as cost, being included in the process, if suitable algorithms can be developed.

Looking at the four best designs, the hull thickness lies in a fairly small range between 99.8 mm and 105.5 mm. The frame spacing however ranges from 544 mm to 745 mm. The additional weight from the extra frames required is balanced by lighter frames, for example the corresponding flange thicknesses are 51.8 mm and 95.2 mm respectively. It is worth noting that although the Pointer method appeared to give good results it took considerably longer to run than other methods.

For the shallow diving hull, Aluminium, with its low density, provided the minimum weight solution. Inspection of the results from both studies showed that no one method could be relied on to provide the absolute minimum weight solution in every case and it is probably wise to use several methods when assessing a new design. This is really no hardship given the speed with which the process can be carried out.

## 3.2 REFINED OPTIMISATION OF PRESSURE HULL USING LINEAR FE

Moving from design equations to an FE model presents the user with a dramatically increased selection of parameters that can be varied. The challenge can be to restrict the selection to a sensible level that will take advantage of this increased scope, but still allow a practical design to be produced.

For this exercise a representative compartment loosely based on realistic scantlings was used as a baseline. The parameters to be varied were collected into logical groups. The frames were separated into two groups, three at each end and four central frames. They were given the same starting values and limits but allowed to evolve separately. The frame spacings, scantlings and shell thicknesses were similarly divided, giving a total of twelve variables, allowing considerable scope for the design to evolve. Figure 3 shows the initial setup (top) and a hypothetical design which would be possible for the given scope (bottom). The mid bay frame spacing was made dependent on the end bay and intermediate bay lengths to maintain a constant overall length.

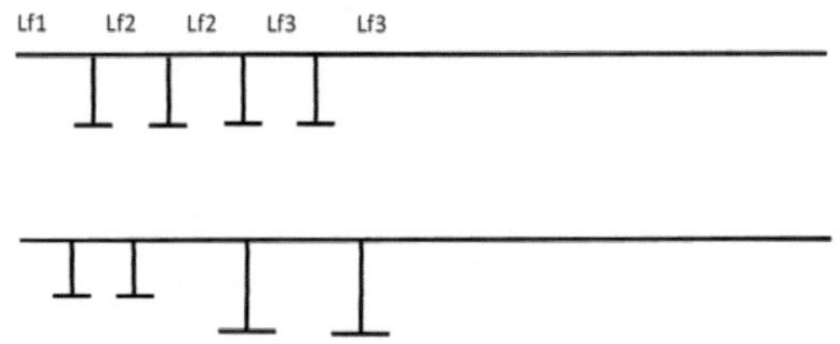


Figure 3. Illustration of the scope of the design space.

Obviously this optimisation took considerably longer than the previous exercise but a study of the various techniques was carried out. The favoured option was the Multi-Objective Particle Swarm method which took about a day to converge on an optimum design after 500 iterations. An initial optimisation gave an 18% reduction in weight but the final frame scantlings did not meet all the design standard stability checks [1]. An additional calculation was added to the workflow to exclude solutions which did not meet these criteria, but still allow complete flexibility in the design space. Despite this restriction, a weight saving of 16% was obtained when the optimisation was repeated.

It was noted that the critical elastic buckling mode of the final optimised structure had switched from interframe to overall and the validity of this design should be verified by other methods.

## 3.3 DOE ANALYSIS OF THE EFFECTS OF SHAPE IMPERFECTION USING NON-LINEAR FE

The elasto-plastic collapse tool was used to set up and execute a design of experiments analysis to look at the effects of the magnitudes of overall Out-of-Circularity (OOC) and interframe imperfections. A modification was made to keep the interframe imperfection as a 3:1 combination of hungry-horse (HH) and the critical buckling mode shape, as described in [4]. A simple calculator component was included to define the amplitude of the HH imperfection as three times that of the buckling mode imperfection, which itself was defined as a variable. Analysis was carried out for 'long' and 'short compartments based on the scantlings of a Notional Compartment mentioned previously.

Some illustrative results are presented in Figure 4 which shows a response surface and Pareto plot for the long compartment (top) and the short compartment (bottom). Perhaps unsurprisingly this shows that the overall OOC has a much more significant effect on the collapse of the long compartment compared to the short compartment. The interframe imperfections are the dominant factor affecting the collapse of the short compartment but are of almost equal significance to the OOC for the long compartment.

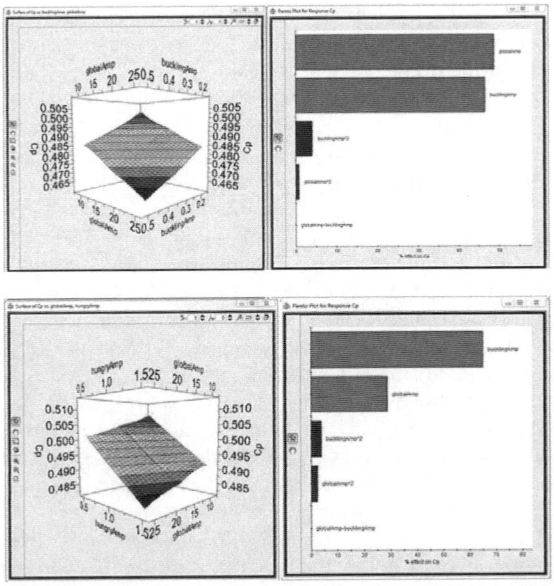

Figure 4. Response surfaces and Pareto plots for 'long' and 'short' compartments.

This initial study was run with only n = 2 OOC. A more rigorous study should consider higher modes. It also used a full factorial approach, which creates and runs a distinct FE model for each combination of the variable. This was feasible as each study required nine FE analyses. This number can increase rapidly with the addition of more variables but *Isight* provides techniques to reduce the number of runs required by appropriate sampling, for example the Optimal Latin Hypercube method.

This case study has demonstrated the ease with which these studies can be set up using the tool and it would be a small stretch to extend its use to more comprehensive and rigorous studies.

## 4. OTHER APPLICATIONS

The *Isight* package also includes software for topology optimisation, TOSCA, and fatigue assessment, fe-safe. Some work has been done to demonstrate the basic capabilities of these tools and this is described in the following sections. These tools undoubtedly have great potential, particularly when integrated into the overall automation and optimisation process.

### 4.1 TOPOLOGY OPTIMISATION

Topology optimisation works by allowing material to be removed from, or redistributed within, a predefined design space. Regions of the design space where material is definitely required, such as load bearing surfaces, can be frozen.

Large hull penetrations, particularly those that intersect frames are an interesting, and perhaps obvious, candidate structure for this type of design. This was demonstrated in a generic test case and Figure 5 shows a snapshot in the sequence of material removal in the topology optimisation.

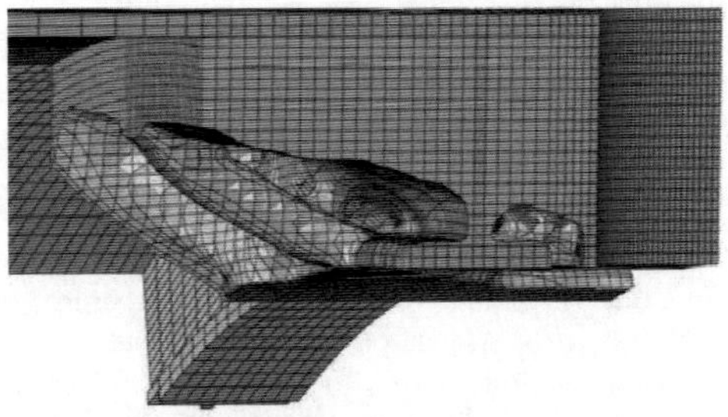

Figure 5. Material removal in a topology optimisation.

This type of topology optimisation usually results in a complex 3-D shape and in the past this would be used as a starting point for refinement to account for fabrication processes. However, the development of additive manufacturing techniques means that it could well be possible to make such components directly.

### 4.2 FATIGUE ASSESSMENT

fe-safe is a suite of fatigue analysis software for finite element models and can be used to calculate:

- Where fatigue cracks will occur
- When fatigue cracks will initiate
- Factors of safety on working stresses
- Probability of survival at different service lives
- Whether cracks will propagate

It can be included as a component within *Isight* workflows allowing fatigue performance to be included in formal optimisation. Although it has not been used in anger as part of the current work, its application has been demonstrated using a simple hull model. Fatigue is an ever-present issue in the operation of submarines and applications for this type of tool are unlikely to be hard to find.

## 5. DISCUSSION

The focus of the work described in this paper has been in demonstrating a new capability and developing some practical applications in the design of submarine pressure hulls and pressure hull components. To that end, the optimisations described have typically used simple objectives and constraints, and relied on default settings for controlling parameters etc.

Even with this restrictive approach, it quickly became clear that simple optimisation routines could produce large amounts of data and a variety of outcomes. Rather than 'dumb down' the process, it is fair to say that an expert eye is still required to interpret the results. The benefit is that demonstrable design improvements have been achieved that just would not have been realised by manual methods.

Work is ongoing in a number of areas: workflows that incorporate legacy analytical design tools for structures other than the pressure hull; and FE model building tools that will allow greater flexibility and control of a wider range of parameters for optimisation and / or sensitivity studies.

## 6. CONCLUSION

The work described in this paper has demonstrated the potential for process automation and optimisation software, and ancillary tools, to enhance the design of submarine structures in particular, and structures in general. It has also exposed some of the difficulties in applying the techniques in practice and described measures taken to mitigate these effects and produce practical design and assessment tools. It then demonstrated these tools in action in a set of case studies

## 7. ACKNOWLEDGEMENTS

This paper describes work carried out by QinetiQ for the MoD under the Maritime Strategic Capabilities Agreement (MSCA) contract. The authors would like to thank their MoD colleagues, particularly Keith Hughes and David Tanner, for their contributions to this work.

## 8. REFERENCES

1. KENDRICK S, W., 'Externally Pressurised Vessels', *in 'The Stress Analysis of Pressure Vessels and Pressure Vessel Components'. Edited by S S Gill, Pergamon Press*, 1970.
2. GRAHAM D. *Predicting the collapse of externally pressurised ring-stiffened cylinders using finite element analysis*. Marine Structures 20 (2007) 202-217. Elsevier.
3. MACKAY, J. R., VAN KEULEN, F. AND SMITH, M. J., 'Quantifying the Accuracy of Numerical Collapse Predictions for the Design of Submarine Pressure Hulls', *Thin-Walled Structures 49 (2011) 145-156*, 2011.
4. GRAHAM D, MURPHY P, HOBSON J, CROSS S, TOOLE C AND FARNWORTH J. *Advanced numerical modelling techniques applied to submarine structures research and development, design, build, and refit and repair.* Warship 2014: Naval Submarines & UUVs. 18-19 June 2014, Bath UK.
5. ISIGHT 2019 User's Guide. © Dassault Systèmes, 2018.

## 9. AUTHORS' BIOGRAPHIES

**Dr Derek Graham** is a Technical Consultant for Submarine Structures and is a QinetiQ Fellow. He has spent the last 30 years at QinetiQ and its predecessor organisations, mostly refining the application of FE analysis to structural issues associated with the pressure hull and other structures.

**Dr Richard Craven** holds the current position of Principal Engineer Marine Composites and Submarine Structures at QinetiQ. He is responsible for leading programmes and providing advice and consultancy to the MOD and other customers on marine composite, and development of tools and techniques for the analysis of submarine structures. He has over 15 years' experience in design and analysis and failure of complex structures and in a previous role at Vestas developed the Blade Analysis Toolkit software suite for the automated design, optimisation and analysis of wind turbine blades.

**Sandra Cross** holds the current position of Senior Structural Engineer at QinetiQ. She has spent more than 20 years at QinetiQ and its predecessor organisations, applying FE analysis to structural issues associated with the pressure hull and other naval structures.

**Michael Lytwyn** is a Senior Structural Integrity Engineer at QinetiQ Rosyth. He has spent over 11 years focussing on the fatigue and fracture assessment of submarine structures. He is involved in both the IMechE and ASME organisations and has published work in several European and International research journals.

**Ali White and Cecily Donaldson** were graduate engineers assigned to the Submarine Structures team at the time this work was carried out.

 

# WHO SAYS THERE ARE NO REAL CHOICES IN SUBMARINE DESIGN?

**D Andrews**, University College London, UK


## SUMMARY

A paper by the author to the last submarine conference emphasised that submarine design is much more constrained than is the case generally for the design of complex surface ships. This is because comparatively the extent of the submarine solution space is usually considerably less and there is the need to go into more detail and achieve a more precise balance for any submarine design option very early in the concept phase. This tighter constraint applies to a greater degree even when compared to designing the most complex surface ships. However, there is a need also to counter the apparent corollary to the above, which might suggest the submarine designer is therefore not faced with a host of major design choices, as should be the case in a properly conducted concept phase for complex vessels. Interestingly, these many remaining choices are both highly interrelated and have to be made very early on, at least in order to achieve the synthesis of any distinct submarine design option. The paper also draws on the more general and extensive exposition presented in the recent Special Edition of the RINA Transactions, by the author, on the sophistication of early stages of design for complex vessels. The current paper therefore points out where designing for submarines is subtlety different to doing so for other complex vessels. This comparison is done before sequentially outlining some twenty significant generic submarine design choices. Finally, the paper points out that that sequential listing has itself, necessarily, by being sequential, thereby further simplified what remains a highly integrated and interactive set of choices.


## 1. INTRODUCTION

The author presented a paper entitled "Submarine design is Not ship design" to the 2017 Submarines Conference (Andrews, 2017). Its message was that submarine design is much more constrained in regards to the extent of the solution space and the need to go into more detail/precision sooner, than is the case for most complex surface ship designs. This had an implied assumption, previously made for surface ship design, that specifically the comparable concept process would be done properly. That is to say that the aim of early stage ship design (ESSD) should be that of Requirement Elucidation, rather than just "a single quick and dirty initial sizing for simple cost analysis" (see Andrews, 2018a).

However this does not mean that the submarine designer is not faced with a host of major design choices, not least because of the need to go into a significant level of detail early in design. This being, in part, because of the more highly tuned nature of a submarine design compared to most surface ships. This is seen to apply even when compared to the extreme of some specialist multi-hulled surface vessels, which could be seen as, possibly, comparably highly tuned as submarines (Andrews, 2016). But interestingly, major submarine design choices also have to be made very early on - at least in regard to each distinct design option, if the project is to be confident that that any specific option can be sensibly costed. Furthermore these choices are basic to assessing a design's technical viability and the programme risks for each potential submarine design option. This then implies there will then be sufficient definition to inform concept level (i.e. high level and strategic) decision-making (see the Third Principle in Elliott & Deasley (2008)).

Thus the paper starts with summarising the unique design features of submarines and why they tend to dominate their design to a greater extent than is the case with most naval vessels. It is this set of demanding engineering issues that were the emphasis behind the 2017 paper on submarine design, that to some degree the theme of the current paper could be seen to be moderating. It is also the case that in comparison with wider service ships, and even more the vast mass of commercial shipping, that still submarines have much in their design environment and processes more in common with other naval vessels, than they are with even other complex ships. To a degree, this commonality, regarding the design process across naval ship/submarine design, has recently been addressed by the author in a major publication (Andrews, 2018a). Thus Section 3 summarises the main aspects of, particularly, the early stages of the design of complex vessels, modifying the generic example illustrations where necessary to those more specific to submarine design issues.

Section 4 consists of the main feature of this paper, which illustrates its contention that there remain many significant choices for the submarine designer to make, despite the apparently restricted design scope, relative to most naval vessels. Thus a sequence of some twenty aspects of design choice is outlined. Theses aspects are presented in a broad pattern from whole boat issues, through discrete design features to some final general aspects, where those choices can themselves have major design, performance and operational impact. The final main section strongly emphasises, through some specific submarine design cases and comparisons that there is artificiality in the previous sequence. This is because, like so many complex designs on a large scale, most of the apparently sequential choices outlined are of course highly interactive with each other and together have to be considered as a complex interdependent and interactive whole.

## 2. THE UNIQUE NATURE OF SUBMARINE DESIGN

The 2017 paper on submarine design commenced by summarising the nature of the design of complex naval vessels in general and readers are referred to that disposition for the full summary. Briefly that paper started with the $S^5$ (Brown & Andrews, 1980) ship design aspects, before addressing combat system related topics, followed by personnel and ship technologies, and then concluding with the unique nature of the warship design process. Regarding the latter topic, it was noted that in this regard the submarine concept process can be just as protracted as that for major surface vessels, not so much due to parallel weapon developments and many diverse solution choices usually typical of naval combatants, but more so arising from the submarine's inherent density of complexity.

This inherent density of complexity applies through out the whole of submarine design and is largely due to the submarine's unique naval architectural aspects, which in summary (from Andrews, 2017) consists of the aspects of: the longitudinal as well as transverse sensitivity for Static Control submerged; Speed and Propulsion in one medium, rather than that for ships on the interface of two; Structural design, where again unlike surface ships the loading (at Deep Diving Depth) is known precisely but the collapse mechanism for the pressure hull is a complex mixture of stress and bucking failure modes; Dynamic Control, especially in the vertical plane where avoiding excursions beyond Deep Diving Depth is a primary boat survival issue.

The 2017 paper went on to address the most critical design issues for submarines (namely Form, Arrangement and Layout; Nuclear Power (if adopted); and Submarine (service) Systems). This was followed by a section on the consequences of conducting submarine concept design inadequately, which addressed: Build and Procurement; controlling Design Development; and why the concept phase differs for submarines. This led on to some concluding remarks on the unique nature of the submarine design process, namely: acquisition issues; support issues; and why "there is no such thing as a repeat submarine design". All this was an attempt to address the subtlety of complex submarine design, be it for large nuclear submarines or ocean going conventionally powered vessels, and these issues will be touched on to some degree in the present paper where the focus is now on design choices.

The current paper addresses an issue that was hard to convey in the previous paper, which first tackled the demanding physical constraints of submarine behaviour and latterly the more strategic issues in submarine design and procurement. Thus it could be argued this paper is trying to both counter the view that there are few submarine design choices (often said to be so because of the first aspect) and that the process at a "technical level" can really be broken down into separate sub-disciplines, rather than the reality of the submarine designer being presented with a fiendishly interrelated set of different design choices. Despite this the twenty or so "aspects of choice" outlined later have had to be considered in a sequence which is neither arbitrary nor wholly realistic in necessarily presenting each as a discrete issue.

## 3. THE ESSENTIAL COMMONALITY IN WARSHIP DESIGN

Having said there are unique choices in submarine design, it is still worth discussing the "submarine variant" of the over riding complex vessel design process that has been outlined recently in a major paper by the author in the RINA Transactions (Andrews, 2018a). This is because these strategic design process issues for such complex systems are still applicable to the design of substantial submarine projects. Furthermore, these issues overlay the more submarine specific "technical" choices, even if very often (as with many naval ship projects) they are inadequately addressed or not even appreciated by the owners/acquisition team/contractors. This is often argued to be necessary in a rush to finish the concept phase, in order to get funding approval and "get on with the "real" design and build". The justification for instead properly conducing the concept phase through requirement elucidation has been argued in that comprehensive 2018 paper (Andrews, 2018a) and readers are strongly advised to read at least Section 3 of that somewhat lengthy exposition, if unconvinced by the brief summary in this section.

In essence the 2018 exposition has three over-riding messages that also apply to submarine design:

- The concept phase for such complex (and physically large) systems is about requirement elucidation of the "wicked problem" or working out both what is wanted and what is affordable/achievable;
- This can only be properly undertaken by designing "inside-out", which means adopting an architectural approach to early stage "ship synthesis" for each design option in exploring the concept boundaries, in both solution and requirements terms;
- Alongside the naval architectural sub disciplines (identified by the first four "$S^5$" aspects (Brown & Andrews, 1980)) there are a host of wider design based issues, which have been categorised as related to Style, for want of a better term.

These messages can be further encapsulated from Andrews (2018a) by three diagrams and two tables, of which Figure 1 and Table 1 of this paper are modified versions of the 2018 ones that reflect more directly their applicability to submarine design, when seen as a (special) variant of complex "ship" design. The original figures and tables from the 2018 paper are:

- Figure 4: "A representation of the overall ship design process emphasizing key decisions" (See detailed explanation of each step in Appendix A of Andrews (2018a)). In Section 5 of this paper this diagram (Figure 1 below) is reproduced (as Figure 4) with the addition of a right hand column showing typical choices appropriate to a specific submarine project. This can be compared with Figure 30 of Andrews (2018a), which exampled a future surface combatant;
- Figure 8: "A Representation of the Solution Space, showing the range of options to be explored". This has not been converted to "submarine practice" as the scope for technology options, both down (utility/commercial practice) and up (unconventional hull forms) is far less extensive in the case of submarine concept studies than is readily possible for surface vessels. Furthermore the other two "dimensions" (of "capability" and "packaging") are, for submarines, limited in scope when compared to surface ship practice, due to the relatively small "payload" impact on the whole boat when compared to that for most surface vessels;
- Table 2: "Three categories of external constraints on complex ship design" which gives examples of each category, most of which also apply directly to submarine design practice as they reflect the naval design environment;
- Table 3: "Types of ship design in terms of Design Novelty." (These range from stretched and "type ship" to radical technologies, which are more akin to aerospace vehicle acquisition – such as for Air Cushion Vehicles (ACV) or Surface Effect Ships (SES).) Most of Table 3 design types are largely inappropriate or not directly appropriate to submarine design practice and so this aspect has not been highlighted here.);
- Table 1: "Listing of Style topics relevant to naval combatant design" (This is reproduced as Table 1 in a modified form here, to reflect submarine design, and can be compared with the original surface ship version, in that most topics are common but two items have been removed and the order of importance has been changed. The two distinctly different submarine topics added are:
  (a) The addition of the importance of ensuring the longitudinal centre of gravity is maintainable, reflecting the extreme sensitivity of a submerged submarine to trimming effects, given the submerged submarine has no waterplane area to resist even slight trimming effects; and
  (b) The addition in the style category of Accommodation to include the necessity for escape from a submerged and disabled submarine, in a manner quite different to and more extensive than the surface ship evacuation arrangements.

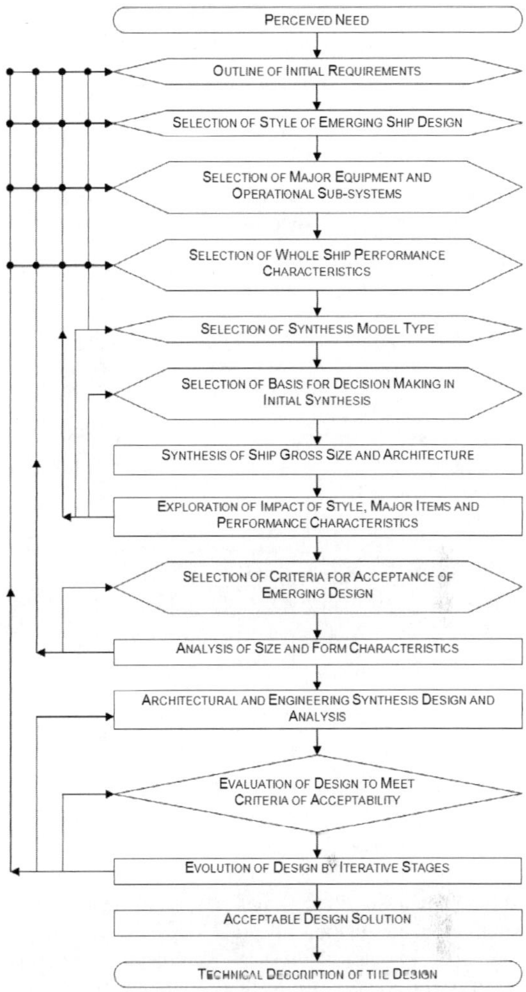


Figure 1: A representation of the overall ship design process emphasizing key decisions
(See Andrews (2018a) Appendix A for explanation of each step and Andrews & Andrews, (2021) for overall picture of this process in a typical ship design environment)

Table 1: Listing of Style topics relevant to submarine design

| **Stealth** | **Protection** | **Human Factors** | **Sustainability** | **Margins** | **Design Style** |
|---|---|---|---|---|---|
| Acoustic signature | Underwater weapon effect | Accommodation & Escape | Mission duration | Space | Robustness |
| Magnetic | Fire | Access | Watches | Weight | Adaptability |
| Infra-red | Shock | Maintenance levels | Stores | Vertical & Longitudinal centre of gravity | Modularity |
| Radar cross section | Damage control | Operation automation | Maintenance cycles | Power | Operational serviceability |
| Visual | Collision | Ergonomics | Refit philosophy | Services | Producibility |
| | Above water Weapon effect | | Upkeep by exchange | Board Margin (future upgrades) | |
| | Corrosion | | | | |

## 4. A SUGGESTED SEQUENCE OF CHOICES AND DECISIONS IN SUBMARINE DESIGN

While Andrews (2018a) has a specific sub-section on submarine design (as indeed it does for Trimaran and SWATH vessels) these are exampled there to show specific differences, for such radical ship types, in their concept phases when compared to designing a "standard" naval combatant. Thus the well-known Burcher and Rydill (1994) outline of the specific concept sizing process for a submarine is partially shown in that paper's Sub-section 8.4 along with Figure 32, which is an extract from Burcher and Rydill (1994) Figure 11.4. However, it is the point of the current paper to emphasise that this 2018 outline and the more extensive one in the 2017 paper, mask the range of decision-making that is or should be involved in *ab initio* submarine design. This was done for a similar reason to the current author's 2017 "submarine design is different" paper (Andrews, 2017), namely, to give a generic overview, whereas the range of examples, given below in the present section, serve to modify that synthesis process outline. This is done by focusing (within the generic pattern of submarine design) on the quite extensive sets of design choices still needed to be exercised in the apparently highly constrained technical design sequence of Burcher and Rydill (1994) or a more exploratory and architecturally based approach for submarine studies undertaken subsequently to Burcher and Rydill's publication but also undertaken at UCL (see Pawling and Andrews, 2011, Purton and Andrews, 2015).

These choices are now outlined from, generally, the most radical or significant in a typical naval submarine design to more specific design features. Some twenty broad features have been chosen to cover the typical set of choices consciously (or, sadly, sometimes unconsciously or unquestioningly) made by a submarine designer. The most important of these are preferably made together with the requirements owner and other significant stakeholders, but should not be overly driven by simplistic initial cost drivers (Andrews, 2018b):

1. Big role choices. In keeping with surface combatants, this ought to be driven by the major mission choices and certainly a Ballistic Missile submarine rather than a Fleet submarine role would lead to distinctly different sizes and costs, if not styles. Yet both these would logically be nuclear propelled. So is the main propulsion choice a binary one of nuclear or non-nuclear? That has been largely the case since USS Nautilus or at least Skipjack and Valiant Classes, for respectively the USN and the RN submarine fleets. However, proposals have been made for hybrid propulsion (i.e. nuclear for endurance with AIP for sprint, see Bradbeer, 2015), which could open up prime mover choices, as might non-nuclear AIP given the potential of future developments. The other current significant role choice, beyond the current Ballistic/Fleet roles, is that of an AUV "mother sub", where the nuclear choice seems marginally secondary to such a radically new role (Purton et al, 2014).

2. After the role choice, which could be argued to be a non-design driven choice, but rather government selection, there are very significant or "Mega Style" choices as discussed in the previous section (see the last column of Table 1). While very unlikely to be as radical as this could be in the case of a surface naval combatant (such as the trimaran variant of the USN Littoral Combatant Ship or an early option of a SWATH study for the UK Future Surface Combatant (Betts, 2018)), there are roughly comparable "big" options, such as the following, which are outlined in order of increasing design novelty:

   - Rebuilding existing submarines with extra pressure hull insert or "SubSLEP" (Ship Life Extension Programme), as was done for the RN post WWII "T" Class SSKs);
   - Follow on of an existing design, which could describe the Trafalgar Class, which commenced straight after the Swiftsure Class;
   - A highly evolutionary design, which some would describe as typical of most submarine design;
   - Radical design, which could range from the distinctly different Swiftsures (to their R.N. predecessors) or the USS Skipjacks (in that this design combined the SSN Skate Class and the hydrodynamically radical (but non-nuclear) Albacore); to having more than one pressure hull (e.g. Dutch triple hulled Zwardfish or the Soviet Oscar SSBN). There are also possible re-arrangements of the "Super Building Block" locations within the pressure hull – thus in some SSN designs the US Navy moved torpedo spaces well away from the fore-ends to acoustically isolate the bow sonar, while the need to incorporate facilities to support swimmer delivery vehicles could make pressure hull and displacement forms quite different.

3. The choice of prime mover(s) beyond the binary nuclear/conventional decision, includes types of diesel electric and electric drive, plus new battery types and configurations, for which current battery R&D may provide distinctly wider choices. The latter will then impact on the "conventional" submarine configuration more markedly than AIP has to date (see design study reported in Andrews et al, 1996).

4. There are big propulsor choices, which are governed by whether the propulsor is driven by power/efficiency or noise reduction (i.e. delayed cavitation onset). This can lead to different types of pump-jet instead of the traditional propeller, the latter being still favoured for conventional submarines. There are possible radical propulsion developments, such as Rim drive (Collins et al, 2015), with major whole boat design implications if such a propulsion systems was to be developed. There are also multiple shaft arrangements and whether to adopt a separate "get you home" or emergency drive system, separate from a main drive.

5. Ballast tankage decisions are interrelated with the choice of external hull envelope shape. Thus the desired level of Reserve of Buoyancy gives the quantification of the Main Ballast Tank (MBT) capacity to provide sufficient freeboard on the surface. This is usually minimised since it is a resistance driver, yet must be sufficient for reasonable seakeeping on the surface, not least for transfer of personnel and stores in extremis. Also relevant is the extent of free flood space outside the MBTs, which has to accommodate the various equipment carried outside the pressure hull, both forward and aft, as well as under the casing. This will include sonar arrays and torpedo tubes (and possible discharge gear) up forward, with any towed array winch and gear aft, plus other equipment that can be external to the pressure hull. The casing height may well be governed by the extent that the forward and aft access/escape towers penetrate above the crown of the pressure hull. There is also, for non-nuclear submarines, the choice of internal or external fuel tanks with issues of compensation and potential of oil pollution informing the choice.

6. The external hull shape is major design choice, which (as Burcher and Rydill (1994) clearly identify) has become dominated by underwater hydrodynamics. The experimental USS Albacore (Jackson, 1987) could be said to have established the Teardrop hull form, constraining the pressure hull and external tankage plus free flood spaces into the essential circular sectioned and stern tapering form with minimum appendages, the largest of which being the fin. Still treated as something of an "add-on" is the large "garage' for a "special ops" swimmer delivery vehicle. Whether this should be better incorporated in the external envelope was explored in a UUV "mother sub" concept at UCL (Purton, 2016).

7. There is limited choice regarding the shape of the pressure hull, which largely means changes through carefully designed transitions in its cylindrical diameter in order to "efficiently" utilise the external envelope. That then, especially if the "teardrop" is very elongated, requires frustrum of cone transitions and implications for where the structural boundaries of pressure hull compartments lie (see Figure 6 in Andrews (2017) taken from Faulkner (1983)). So the desire to keep down the overall (hydrodynamically significant) envelope volume can then complicate the already very heavy pressure hull structure. Such complicated structural details can then have implications for the hull's resistance to underwater

shock, since complicated pressure hull transitions can introduce stress concentrations and poorly responsive "hard points" (see Andrews (2017) Figure 9). Conversely an almost constant pressure hull diameter, as introduced by the R.N. Swiftsure design, see Figure 5 in Andrews (2017) and, for an SSK equivalent, Figure 6 in Wrobel (1985), leads to large torispherical end closures (dome bulkheads). These have their own issues of sensitivity to flatness and care required in designing and necessitating complex welding of penetrations, plus the implications of "hanging" the forward and aft external structure off each dome bulkhead.

8. Beyond the external aft end structure, most designs have a central propulsion shaft leading to a stern arrangement forward of the propulsor where again there are design choices. Thus the upper and lower rudder and port and starboard aft hydroplanes in the traditional cruciform arrangement can be modified into "X", "Y" or "inverted Y" as Figure 7 of Andrews (2017) suggested. These options can be complicated by different propulsor types and by other features, such as dispensing and retracting towed array sonar cables – as is shown very prominently in some Russian SSNs.

9. There is a clear desire to keep to minimum the number of big pressure hull penetrations, as a primary source of submarine pressure hull failure (such as in the sinking of USS Thresher in 1963), plus major penetrations add to the already large proportion of displacement due to the steel weight of a submarine, in comparison to surface vessels. Other penetrations like escape towers, logistic accesses and diver accesses tend to be combined to reduce to a minimum of two, one each forward and aft of the fin. Similarly masts and periscopes are kept to a minimum and collocated in the fin, while the mass of electrical penetrations, such as for sonar arrays tend to be under the casing and penetrate the pressure hull through a few thickened inserts. There are also the operating gear for 'planes and rudders, major machinery inlets and outlets for nuclear secondary propulsion plant, plus the various "hard" seawater systems and other penetrations, such as communications fits, decoy dispensers and a gash ejector. So the desire to minimise the extent and impact of so many openings can influence system configurations and hence the wider design.

10. There are choices with the casing and external keel (if the latter is fitted, as often is the case with SSKs, since they are more likely to operate where there are relatively shallow continental shelves and seaways, such as the Mediterranean and the Baltic Seas). Such keels are also useful for stowing the variable solid ballast, as so low down but such projections are not hydrodynamically attractive in highly manoeuvrable SSNs. The casing's size tends to be driven by sufficient height for ready access but not excessive height and the adoption of GRP panels, for ease of access in harbour and reduced hull maintenance, is often attractive.

11. The common locations chosen for the forward hydroplanes reveal a typical submarine design conflict. Thus many designs (such as in the US Navy) place the 'planes on the "sail" or fin while recent UK designs have an "eye brow" arrangement well forward and with the operating mechanism in the forward external structure or casing. Neither of these are that hydrodynamically efficient in controlling vertical attitude, particularly at low speeds, as this would require the forward 'planes to be on the centre line and as far forward as possible. In fact the USN arrangement results in very large forward 'planes, but all this is due to prioritising the overriding operational need to keep the flow and mechanism of the forward 'planes' as far as possible from the forward facing sonar fit, with consequently very large forward 'planes There is a choice (at least for small conventional submarines) of retracting the forward planes into the free flood space when not required at higher speeds, however issues of reliability and space demands have usually questioned adoption of this added complexity.

12. The level of automation can have a major implication beyond safety and cost issues in that the number of personnel on board (even with the very cramped living arrangements on all submarines compare to even naval surface ships) is a major size determinant. This issue, is driven by the "manning" practices of individual navies. Thus certain navies, such as the Norwegian Navy, seem to man their submarines very leanly, however this can be a combination of organisational structure (just three levels of management rather than that of four levels in bigger navies), greater levels of automation and requiring a high skill set (all the crew are graduates). Such an approach may well become the way forward generally but needs more than just an individual submarine programme's design decision, as "manning" is a navy wide issue and culturally driven.

13. The location of the Control Room (CR), as the most important manned space, is key at least to the layout of the pressure hull forward of the main machinery spaces. Traditionally the Control Room was placed under the fin and on the top deck so the periscopes could have maximum reach above the hull at periscope depth and be of maximum length when retracted down to the keel. However submarines can now use electronic imagery so fully pressure hull penetrating periscopic masts can be replaced by optronics, which implies the CR can be both on a lower deck and not directly under the fin. There remains a good reason for the CR being within easy

reach of the forward end of the fin and that is that the conning tower access to the bridge and back to the CR when surface running is important. Nevertheless in theory there is much more freedom as to where the CR can be. One major installation issue can be eased once the CR is no longer on the pressure hull upper most deck. With the CR at mid height in the pressure hull no longer is the CR layout confined to the centre of the space transversely which results in PH curvature leaving little useful space for people and equipment, beyond pipes and cables. It is even possible to adopt "a floating deck" for the CR and associated electronic offices with cabling and cooling systems under the CR floor, which can be installed just before the pressure hull units are closed up prior to launch.

14. The height, length and maximum width as well (now) the location longitudinally of the fin is still driven by the need for the number of telescopic masts which provide various radar, EW and, now, electronic periscopes, as well as the snort arrangement when running diesels at periscope depth for both propulsion and recharging the batteries. Then there is the bridge for surface conning and the possibility of non-torpedo like weapons such as anti helicopter missile fits that have been proposed, as have modular weapon fits extending the fin or inserted into the pressure hull (Leadmon et al. 2002). Given the fin is the largest appendage to the hydrodynamically efficient external hull, its extent is usually kept to an absolute minimum and its shape, particularly where it meets the casing, further hydrodynamically refined. If a large "hangar" for a swimmer delivery vehicle is required then this is usually located on the casing aft of the fin but this then further adds to the appendage challenge.

15. The major sonar fits, whether passive and integrated into the external hull/casing or reeled out of the stern from a winch in the aft external structure, as a towed array, can be overriding location drivers. As indeed can the active bow array taking precedence over torpedo tubes, forward hydroplanes and much else. Thus the bow fit of the array of transducers can drive the torpedo tubes to lower locations than desirable for reload evolutions through a further penetration at the crown of the pressure hull. If the fin also locates the forward hydroplanes far from the most hydrodynamic efficient location, as previously mentioned its arrangements are further complicated, which is far from ideal for such a safety feature.

16. Given that acoustic signature is key to underwater performance of any military submarine, it should dominate many of the choices the submarine designer makes. Despite this it is often thought that underwater radiated noise (URN) performance is solely relevant to the detailed design phase. However, as Hamson (2016) points out many of the key trade-offs in the Concept Phase are critical to minimising URN. Hamson lists:

- Hull size, since it drives resistance and hence propulsive power, so being the major energy source;
- Structural arrangements, since structure borne noise with a high dynamic impedance is magnified given the pressure hull's mass, especially if equipment sourced noise is not adequately isolated from the hull and hence transmitted via structure to the sea;
- General arrangement, governing many the potential noise transmission paths;
- Propulsion machinery may be chosen for its URN performance or ease of reducing the latter, even at the cost of slightly reduced propulsive effectiveness. This can extend from keeping the shaft speed as low as possible, often through a big reduction gearbox or an electric motor, to the propulsor being a pump jet rather than the traditional propeller, both measures being adopted to prevent early onset of blade cavitation at higher speeds;
- Whole boat trade-offs, where all the above must be brought together taking account of the interactions. This can almost be regarded as the epitome of the highly interactive nature of submarine design. Hamson highlights the "significance of 'weight and space budgets' for noise control engineering" and shows a plot that (for a typical SSK) indicates larger hull sizes can be detrimental to URN performance across the range of significant frequency bands.

The extra whole boat URN reduction features were a major determinant, alongside other primary drivers, in adding extra frame spacings in the Trafalgar Class over their predecessor, thus enlarging the boat's length but not the diameter of the pressure hull.

17. Accommodation choices are part of general arrangement choices but also crucial to easing "working the boat" and also the need to reflect provision for future levels of automation. It is important to avoid the Control Room being on the main forward and aft access in larger submarines (SSN/SSBN). This can be facilitated by adopting electronic periscopes but moving the Control Room away from being under the fin is then at the price of the bridge being further from the Control Room than might be sensible.

18. Distributed Ship Service Systems (DS3) are often seen as the Cinderella aspect for consideration in early stage submarine design "as one needs too much definition to consider in ESD". Yet the modern submarine is fundamentally dependent on its many and diverse distributed systems, integral to safety and operability. Without a means to explore

different DS3 arrangements in ESD submarine design is rapidly constrained to adopting existing solutions regardless of potentially better alternatives. This need for early stage exploration was the major driver behind current research into submarine DS3 in ESD at UCL (Mukti et al, 2021).

19. Construction issues. Despite the level of outfit congestion due to the perceived "cost" of meeting the weight and buoyancy balance, any submarine design must be buildable. Issues such as accommodating floating decks, for ease pre-outfitting before pressure hull units are closed up, and special fabrication facilities, for welding high strengthen pressure hull steels, need early design decisions to be incorporated. For nuclear powered vessels the demanding nature of installing nuclear plant and the added expense of meeting the concomitant safety standards can be seen in the size of SSNs and their enormous cost to build, particularly if the extent of the special facilities required is added in. Strangely, the impact of building submarines is not always solely a question of technical issues (fabrication, installation, launch and setting to work) but sometimes political and then consequential additional design implications. This occurred with the proposal by Kockums for the Australian Collins Class, where that contractor's winning feature was seen to be to "build a major part of each boat in each major Australian state", rather than specific operational and technical features (see Section 6.1 of Andrews (2017) for more extensive discussion of the Collins Class acquisition history).

20. Maintenance Issues. The importance of correctly managing this aspect of submarine ownership has been particularly reinforced by the Coles Report (Coles, 2012). This demonstrated that to improve the availability of the six Australian Collins Class SSKs to that expected by leading submarine fleets (namely 50% of the fleet being available at all times) required addressing five root causes. Some 25 specific recommendations, many of which were organisational, were proposed to achieve this necessary improvement for the Australian fleet. Thus TL Support must be part of the project from the start and where features are likely to enhance reliability, sustainability and availability they have to be given due weight, even though that usually means a larger hull volume than crude weight and UPC minimisation driven designs "allow". What has often been regarded as something to be addressed late in design must be given adequate space, weight and margin disposition from early on to enable the acquisition team to adequately address supportability. Inadequate support provision in a design can even lead to shortening the life of that class – so consequences of insufficient consideration can waste financial commitments out of all proportion. Apparent details, such as LETs or a Dutch Breech (Wrobel, 1985) need to be decided early on, given the likely impact on pressure hull structure and General Arrangements (e.g. adequate removal routes).

## 5. COMMENTS ON DECISION LISTING

The first point to make is to, once more, emphasise that the "20 Items" in Section 4 may seem to be a sequential listing of discrete topics but of course they are not. As with the exposition in the paper on submarine design four years ago to the last Submarine Conference (Andrews, 2017) and the classic textbook by Burcher and Rydill (1994), any outline of the various aspects of something as so interdependent and integrated as a submarine design, when necessarily presented in discrete topics, is in danger of conveying another wrong message. Almost all the listed topics interact and the integration of them interacting with each other and on the total submarine design balance cannot be over emphasised. Thus in the 2017 paper, the order in which the main naval architectural topics, as they are sequentially mentioned in the introductory section of this paper, is different to that chosen by Burcher and Rydill for their seminal book. It is also different in order and sub-division to the listing in the last section. The latter tries to follow a pattern of very major decisions early on but clearly some of the later more transversal issues, like noise reduction, could be argued (as the main design driver for operational effectiveness) to be earlier listed, or does overall design balance and safety still trump noise signature as the key submarine design issue?

It is very hard to convey the degree of interaction, interdependency and integration between all 20 topics covered. In expositions of general (complex) ship design, simple interdependency or interaction diagrams are sometimes used to show how various design decisions emerge. Thus Rawson and Tupper (1976) show in their chapter on ship design at Figures 15.9 and 15.10, two interaction (or design influence) diagrams, one for the Weather Deck of a destroyer and the second a more generic diagram for how the initial and final arrangements of the ship's set of fuel tanks might be decided. The current author showed how a novel ship hull form, that of the trimaran, had interdependent features such that if one changed then the design must go round an iteration until the "knock on effects" across several features have been absorbed by the iterated (and enlarged) design (see Figure 2 from Andrews, 1994). This example was relatively simple but if you take a SSN example of the interdependencies of just part of a submarine design, as shown by Collins et al. (2015), reproduced at Figure 3, then the ability to represent the interactive consequences of a change in just one aspect looks formidable.

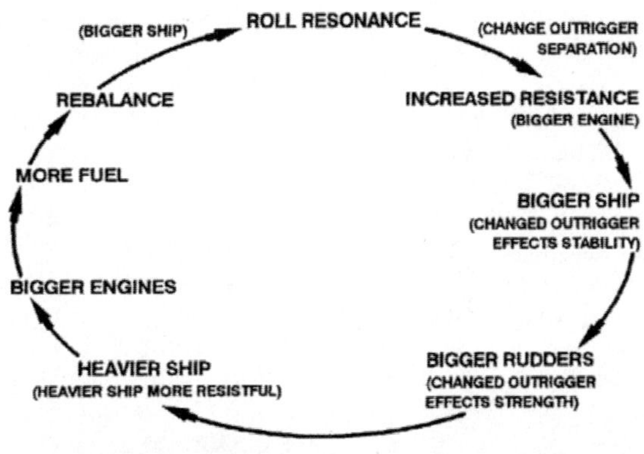


Figure 2: Trimaran example of design interaction due to an imbalance requiring iteration (Andrews, 1994)

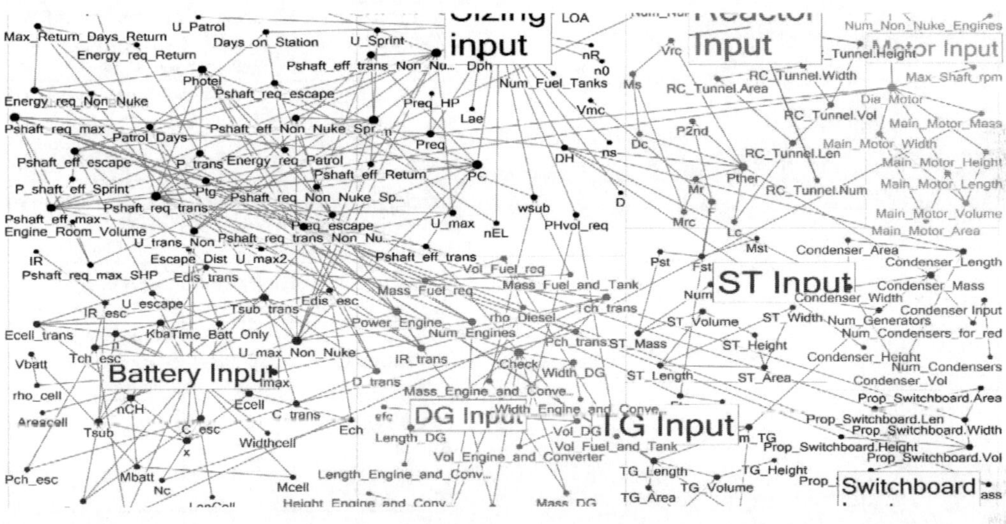


Figure 3: A representation of naval vessel complexity – a partial network model of a nuclear submarine's systems structure (Collins et al 2015)

Figure 4 takes the major design decisions, including the choices designers consciously, or even un-realisingly, make regarding the wider design environment indicated by Figure 1, and puts examples alongside each step that might apply to a fairly straight forward SSK design. This is similar to that added to the same figure for a typical frigate (see Figure 30 in Andrews, 2018a). The comparison of these two examples is quite informative in that the submarine case can be seen to be more specific or even more readily convergent on a solution, given the need to confirm early selection of features to achieve a much tighter design balance than that which is usually necessary in the Concept Phase for a surface vessel, combatant or otherwise.

However it could still be argued that the decisions that are made depend on quite subtle issues, such as the extent of margins adopted in departing from previous designs. This can be seen in the comparison between two then concurrent designs that of the UK SSNOX design, which became the Swiftsure Class, and the USN Skipjack Class. The former was a radical departure from its predecessor class of the Valiant/"C" boats, as already indicated in Item 2 of Section 4 above, having no pressure hull transitions and almost fixed diameter pressure hull, while the latter perpetuated the Albacore teardrop form with a long tapering aft form. This could be seen as a compromise in the former case for flow into the large pump-jet to delay cavitation onset, while the latter was about obtaining a high top speed.

However it is unlikely that the Swiftsure design, in giving greater internal volume, particularly aft, was initially seen as primarily a design providing ample growth potential or adaptability. Yet this was fully exploited by the follow-on design of the Trafalgar Class, which had both more space but more importantly adequate stability margins, both vertically and longitudinally, to take advantage of the earlier design's margins. The original SSNOX project director, Norman Hancock, was apparently obsessed with controlling weight growth as the design developed, such that the margins provided the ability to readily achieve balance in the follow-on design. It sounds simple in hindsight but the example belies the usual difficulty in "repeat submarine designs" (see Section 6.3 of Andrews (2018a).

A conventional submarine comparison, which further points out the subtlety of submarine design, is the comparison between the UK 2400 design (Upholder Class) and the Kockums originated Collins Class for Australia. It might be thought that these two would be comparable but not only do they exhibit distinctly different styles but also originated from needing to meet distinctly different roles. The former was a UK Cold War ASW submarine designed to operate very specifically in the Greenland-Iceland-UK (GRIUK) Gap to "sit very quietly" and catch Soviet SSNs trying to break out south into the Atlantic, while the latter were a direct development of Swedish SSKs with short endurance consistent with operating in the Baltic against a threat from the Soviet Baltic Fleet. Both designs retain features from their predecessors and exemplify the importance of both the early choice of "Selection of Style" (see Figure 1 and detailed issues in Table 1) and the complexity of the design environmental factors that Andrews and Andrews (2021) illustrate in their "Rich Picture" of the naval ship design process.

**PROCESS**

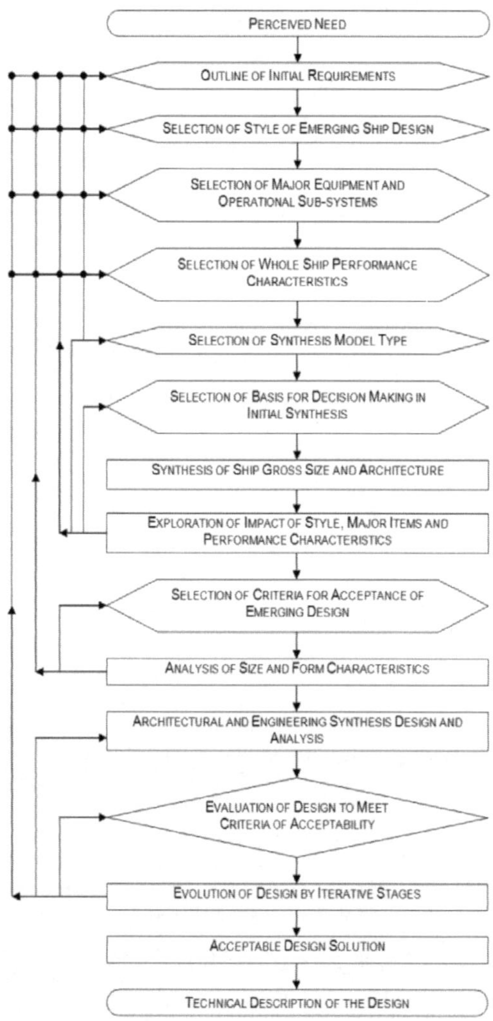


**EXAMPLE**

**A Future SSK**

| |
|---|
| **4 Torpedo Tubes 20 reloads, Towed Array, ASW, 275m DDD, 16kts sprint/7kts snort speed, IR <20%, single submarine communications fit, etc.** |
| **Macro Level Style: Conventional Single pressure-hull with fin/casing; Main Level: Current practice – Table 1** |
| **Propulsion: IFEP, Systems: Standard redundancy** |
| **$S^4$ (really Submarine Stability + top speed/battery/snort endurance) + Some specific Style sub-elements/standards)** |
| **DBB approach using Paramarine as exampled in Mukti et al. (2021) with Network Flow for DS3** |
| **UPC while meeting simple sub balance criteria (Vert&Longit Stability informed by DBB derived KG + top speed/battery/ snort endurance)** |
| **Actual synthesis summarised by Mukti et al (2021) example.** |
| ***See note below*** |
| **This is not just cost but also the relevant submarine engineering aspects, especially mission specific aspects** |
| **This depends on how far this option is to be taken – it is likely only one or two options proceed further** |
| **The extent of layout work up (in this case including DS3) and interaction with engineering analysis depends on both apparent design drivers and the higher level of NA/engineering balance merging into Feasibility and beyond.** |
| **Evaluation should now be extensive and associated with emerging detail, such as structural approval and eventual safety assurance.** |
| **The last three boxes describe the philosophy behind the total submarine design process well beyond concept design of the chosen option.** |

Figure 4: A partial representation of ESSD for a SSK example when in support of Requirement Elucidation with example steps from Figure 1

*Note to Step 9: Exploration would be limited for a given design option to exploring a narrow range of performance and payload variations where need to achieve tight weight/buoyancy/ vertical & longitudinal balance is critical. The fuller exploration, following a Figure 8 (in Andrews (2018a)) like exploration, with a wide extent of design choices, is less typical for submarine exploration where exploring the design process novelty level is less likely than for surface vessels (see Section 5 of Andrews (2018a)).*

 

It can be concluded that emphasising the need to achieve design balance is ultimately the characteristic that is key to submarine design. That balance is ubiquitous but ultimately boils down to balancing the over-ridding safety concerns (given the catastrophic consequence of any major feature failing and the submarine sinking well below DDD and the pressure hull explosively collapsing) against the need to adopt sufficient novelty to improve the odds against likely opposing submarine adversaries. The dependency on slight margins of capability will still apply to SSKs and SSNs in one to one conflicts, while these no longer seem as likely for surface combatants. The latter been seen as operating as parts of network centric warfare due to the ubiquity of electronically "visible" surface/air warfare, while the submarine still has that edge of being the silent service.

Predicting the future of submarine design must be as problematic as any other crystal ball gazing, but certain issues look to be part of the mix:

- A wider set of roles, such as "UXV mothersubs" (Binns et al, 2011);
- More propulsion choices from a growing range of AIP solutions for smaller "conventional" submarines to hybrid and even fusion reactors for Fleet and Ballistic Submarines (Warship Technology, 2015);
- Increasing levels of automation with the promise of smaller submarines due to smaller crew numbers;
- More integrated designs that emphasises through life supportability (Coles, 2015) including adaptability;
- Better stealth features across a wider spectrum of signatures;
- Retaining the recognition that submarine safety and sub-systems reliability are paramount in a balanced design.

There are of course other parties than the naval architecturally trained submarine designer involved in the above choices and in that sense again simplifications have had to be made. Also for some aspects, such as acoustic signatures and sonar (and other sensor/weapon sub systems), developments and demands on the design choices cannot be readily discussed in the open literature. However the range of choices above are many of the most crucial that need to be taken into account and have been listed to moderate the feeling often implied that submarine design doesn't just differ in the "tighter" demands of the naval architectural sub-disciplines over that of most surface vessels, but means the designer has limited choices. This paper has set out to counter that view and encourage a more exploratory approach to early stage submarine design.

## 6. CONCLUSIONS

Submarine design is both distinct and has much in common with the overarching process for designing complex vessels, particularly in the earliest crucial design stages. However, the scope for choice in the case of submarines is heavily constrained by the need for balance, particularly in overall weight and buoyancy plus longitudinal as well as vertical sensitivity, to a much greater degree than that is necessary for most surface vessels. Nevertheless there are choices to be made and as with surface vessels the message is that those choices can only be made very early in the design and through a relatively wide and considered exploration of the impact of any possibly significant choice.

The sensitivity of any submarine design study is such that that needs to be done in sufficient detail to ensure there is adequate balance in the design to allow for the development of the design in detail. For unlike a surface vessel, the design team cannot readily add additional superstructure or accept a little more draught, as space and weight growth may demand, as is too often the case in the post-concept phases of surface vessel projects.

## 7. REFERENCES


1. ANDREWS, D J: (2017), *Submarine design is Not ship design*" RINA Submarines Conference Bath, June 2017.

2. ANDREWS, D J: (2018a), *The Sophistication of Early Stage Design for Complex Vessels*, Special Edition IJME, Vol. 160.

3. ANDREWS, D J: (2016), *Unconventional Ships and Unconventional Design*, IIPER Conference, Cortona, Italy, Oct 2016.

4. ELLIOTT, C and DEASLEY, P:(2007), *Creating systems that work: Principles of engineering systems for 21st century*, Royal Academy of Engineering, London, June 2007.

5. BROWN, D. K. and ANDREWS, D J:(1980), *The Design of Cheap Warships*, Proc. of International Naval Technology Expo 80, Rotterdam, June 1980. (Reprinted in Journal of Naval Science April 1981)

6. ANDREWS, D J and ANDREWS, D A W: (2021), *A new way to visualise the design process of complex vessels,* The Naval Architect, Jan 2021

7. BURCHER, R. and RYDILL, L J:(1994), *Concepts in Submarine Design*, Cambridge UP, 1994.

8. PAWLING R. and ANDREWS, D. (2011a), *A Submarine Concept design – The Submarine as an UXV Mothership*, Warship 2011: Submarines and UUVs', RINA, Bath, June 2011.

9. PURTON I.M., & ANDREWS, D.J. (2015), *The use of Computer Tools in Early Stage Design*

*Concept Exploration to Explore a Novel Submarine Concept*', Proc 12th IMDC, Tokyo Univ, May 2015.

10. ANDREWS, D J :(2018b), *The Conflict between Ship Design and Procurement Policies*, RINA Warships Conference, Bath June 2018.

11. BRADBEER, N (2015), *Design of Submarines for Hybrid Nuclear/Battery Propulsion*, Proc 12th IMDC, Tokyo Univ, May 2015.

12. PURTON, I M et al (2014), *Computer Simulated Trends in Designing A Future Unmanned Underwater Vehicle 'Mothership' Submarine*, RINA Warships: Submarines & UUVs, Bath, June 2014.

13. BETTS, C V :(2018), *30 Years of Naval Construction 1983-2013*, RINA London 2018

14. ANDREWS D J et al. (1996), *SUBCON - A New Approach to Submarine Concept Design*, Warships 96 'Submarines 5', RINA, London, June 1996.

15. COLLINS, L et al. (2015), *A new approach for the Incorporation of Radical Technologies: Rim Drive for Large Submarines*, Proc 12th IMDC, Tokyo Univ., May 2015.

16. JACKSON, H (1987), *USS ALBACORE*, RINA Submarines Conference, London, June 1987

17. PURTON, I M (2016), *Concept Exploration of a Novel Submarine Using Innovative Computer-Based Research Approaches and Tools*, PhD, UCL 2016

18. FAULKNER, D :(1983), *The Collapse Strength and Design of Submarines*, RINA Submarines Conference London, June 1983.

19. WROBEL, P :(1985), *Design of the Type 2400 Patrol Class Submarine*, Trans RINA, Vol. 127, 1985.

20. LEADMON, J T, et al.:(2002), *Submarine Payload Modularity for an Uncertain Future*, RINA Naval Submarines Conference, London, June 2002.

21. HAMSON, K G :(2016), *A New Approach to Managing the Acoustic Signature Risk in Warship Design*, RINA Warships Conference Bath, June, 2016

22. MUKTI, M H et al (2021), *Distributed Ship Service Systems Architecture in the Early Stages of Physically Large and Complex Products*, IJME 2021tbp.

23. COLES, J :(2012), *Study into the Business of Sustaining Australia's strategic COLLINS Class Submarine Capability*, Commonwealth of Australia, Canberra, Nov 2012

24. RAWSON, K J and TUPPER, E C :(1976), *Basic Ship Theory*, Longman, London, 1976.

25. ANDREWS, D :(1994), *Trireme to trimaran: the fascination of ship design*, Inaugural Lecture, UCL, Sept 1994.

26. BINNS, S D, et al.:(2011), *A Vision for an MXV and UXV enabled future Host Submarine*, RINA Warships: Submarines and UUVs, Bath, June 2011.

27. WARSHIP TECHNOLOGY :(2014), *Compact fusion reactor could power next-generation warships*, RINA, Jan 2015.

**David Andrews FREng** was given a new Chair in Engineering Design at UCL in 2000, following his early retirement, when a Director (1*), from the UK Ministry of Defence. His early career in the RCNC included several postings in both in-service conventional and new construction nuclear submarine design sections. He was subsequently Head of Preliminary Design responsible for all initial naval design studies, including on the Future Attack Submarine. From 1993 to 1998 he was appointed the MoD Professor of Naval Architecture at UCL where he directed the MSc in Naval Architecture and the post-MSc Submarine Design Course. In 2001 he set up a new Design Research Centre in the Department of Mechanical Engineering, which is focused on CASD and the design of physically large and complex systems. He is a Fellow of RINA, a former Vice President and currently Editor-in-Chief of the IJME. In 2000 he was elected a Fellow of the Royal Academy of Engineering and is currently the International Chair of the triannual International Marine Design Conferences. In 2020 he was awarded RINA's highest award, the William Froude Medal, "for his contribution to the field of ship design".

# CAD SYSTEM DISTINCTIVE FEATURES IN SUBMARINE DESIGN

**R Perez-Fernandez,** Universidad Politécnica de Madrid, Spain

## SUMMARY

Although the object of surface ships and submarine industries is the same (a floating artefact aimed to comply with a particular function), the particularities of this same function, the ways for achieving it, the costs involved, and the metrics used to analyse results are so different that in fact make both industries different, usually considering that tools and processes used in both should also be different. This includes CAD Systems, although in this case the dichotomy was not established in the very beginning, but formed as the processes in submarines and surface ships diverged with the pass of the time. In fact, CAD Systems were first conceived for solving problems in other sectors like aerospace/defence, as these industries had the resources to invest in such advanced technology. Submarines has been often perceived as a product with more similarities with hi-tech industries than with traditional shipbuilding. However, the impressive growth of surface ships and its requirement for reducing costs and delivery time made shipbuilding CAD Systems suppliers focus their attention in it as a source for obtaining incomes, developing, and adapting systems to this particular industry. General CAD Systems took advantage of the situation to implement themselves in submarine, where they remained (still remain) for many years. But the requirements for reducing costs also catch up with submarine, so in many cases they decided to put forward the replacement of general purpose CAD Systems with shipbuilding specific ones, and this produced an automatic reaction of shipbuilding CAD Systems suppliers that have devoted more effort to try to solve particular submarine problems. In order to check the convenience of the use in submarine of shipbuilding CAD Systems, nowadays focused in surface ships, first of all it should be studied the differences between both surface ships and submarines. Once established them, we would be in position to propose a suitable solution. As summary, this paper covers the differences between surface ships and submarine as well as provide a series of recommendations toward the design of current and future CAD Systems.

## NOMENCLATURE

| | |
|---|---|
| CAD | Computer Aided Design |
| DM | Documentation Management |
| PLM | Product Lifecycle Management |

## 1. A BRIEF HISTORY OF CAD SYSTEMS

In the late 19th and early 20th centuries the frames of a steel ship were stood up on the keel like those of a wooden ship, and the plates attached later. Frames had to be shaped to match the curves of the hull design. Each one was heated in a forge and then hammered or jacked to match the shape of its template. In late 19th century, it was not easy fairing the hull. Flexible sticks, called battens, were used for fairing the lines, i.e. checking the measurements against the lofting and making sure they looked fair with no kinks or irregularities. Finally, the lines were transferred onto full-sized moulds (patters) sawn from thin wood. These were traced to shape the timbers for the ships, much as a dressmaker uses paper patterns to cut cloth for a garment. Loftsmen cut the mould stock (thin pine boards) to shape using a bandsaw. Each pattern was marked to indicate which piece of the frame it was. The patterns could be stencilled with the hull number to keep track of the sets. Often two or three sets of moulds were made: one to mark and cut the timbers, and one or two up in case the originals were damaged. Ship designers in the design office drew up construction details and compiled tables of offsets, or measurements, for the full-scale hulls. At some shipyards, drawn plans were used instead of, or in addition to, a half-model. In some shipyards, carving a half-hull model was the first step in designing a new boat (Perez & Gonzalez, 2015).

Figure 1: The "Peral" on the slipway of La Carraca dockyard, 1888.
*Source: www.elespanol.com*

Drawing on both tradition and experience, shipbuilders carved hull forms that had been proven seaworthy and economical. Half-models were carved from layers (or lifts) of wood. Sometimes contrasting wood types indicated design features like the waterline. Old shipyards used models with the lifts glued together, as were used during the construction of the *Peral* (illustrated in *figure 1*), taking measurements from the outside of the model. Throughout the first half of the 20th century, ships were getting bigger; so it was necessary to work on larger scales. The templates allowed working on different scales, such as widely used 1:10. However, with growing ship size, the moment came it was no longer practical to use templates. This happened at a time when the first computers came to our industry, promoting the development of ship design *Computer Aided Design*

(CAD) systems (Perez & Peter, 2019) that started dealing with the hull forms fairing, from where the hull structure would also be generated. Outfitting handled by CAD systems (pipes, machinery, HVAC...) is considerably newer than hull form design or structural design. Outfitters used wooden and plastic models until the late 20th century. It was only then that CAD systems expanded the application range, covering all project disciplines, thanks to the evolution of computer graphics environment, among other key factors.

Traditionally, most shipbuilding CAD systems focused on hull forms definition, naval architecture calculations and structural design. This changed when new challenges in shipbuilding forced marine suppliers to devote special attention to this matter, demanding closer coordination between hull structure and outfitting. It was a time and cost saving matter, and most shipyards started to pre-outfit the units before their assembling in the slipway.

Some marine CAD systems started the development of their outfitting tools, others preferred to find a closer integration with existing plant design-oriented systems. The development of a particular outfitting design tool was based on the fact that the actual requirements for outfitting design are not limited to a close integration with the structural design. Problems to be solved, regulations, working procedures, nomenclature, production information, etc. are so particular to shipyard environment that it is convenient to have a dedicated tool rather than try to adapt an existing one.

As time went by, outfitting tools have been increasing the scope of support. Today, tools usually include specific environments for equipment modelling and layout, piping and HVAC ducts routing, definition of auxiliary structures (foundations, gratings, ladders ...), and definition of distributor supports and hangers.

In some cases, also electrical and accommodation aspects are considered. Singularities of outfitting design require to work in a pure 3D environment and with a friendly and suitable user interface, but new developments in outfitting tools have been always handicapped by the available technology (hardware, graphic possibilities...).

Nowadays it is commonly assumed that outfitting tools should be able to work in a solid visualization method, with huge amounts of information on the scene being dynamically handled (Perez & Gonzalez, 2015).

## 2. INTRODUCTION TO THE INDUSTRY 4.0

After analysing the past in previous chapter, it is highly recommended to talk about future in shipbuilding, mainly because submarines are comparable with spacecrafts, and any new technology may be applicable first in this type of extremely high-tech product. The Industry 4.0 transformation implies a huge collection of interrelated and joined technologies, which can be analysed independently, but must be applied as a complete integrated implementation in each industrial field, moreover, in the CAD development industry. Each technology exposes a set of boundaries, which cannot be differentiated evidently from its neighbouring technology.

Virtual, Augmented and Mixed Reality are closely associated to the Digital Twin and interconnected with the Big Data, which is produced by the CAD and *Product Lifecycle Management* (PLM) tools and all surrounding solutions, which applies some cloud/edge/fog computing to this data in a merged technology between finite state machines and *Artificial Intelligence* cognitive processes (Benayas and Perez, 2019).

To perform in an agile manner all these computing, it requires a network which supports different connection ways to add special devices, i.e. *Internet of Things* (IoT), which can access to the data, creating and modifying it, in a different layer which affects to the basic information layer created by the CAD System in the shipyard. This network should be safe, pertaining cybersecurity, but open to allow distributed work, which must be step controlled in a manner that records any modification of each working step done in an open, transparent, trusted and non-modifiable working method for all actors involved in process, like the shipyard, the design offices, the classification society and the navy, blockchain. Moreover, blockchain technology can address the issue of security (Perez et al., 2018).

Results of the design should be easily integrated with future building ways like 3D printing, generating printing orders directly from the CAD model. Shipbuilding engineering phases involve design and production, but an integrated Industry 4.0 CAD System can play an interesting role, involved in operation and maintenance.

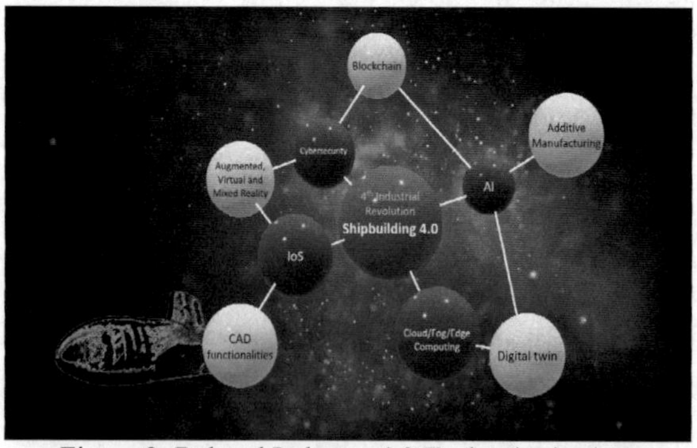


Figure 2: Related Industry 4.0 Technologies in a shipbuilding CAD environment

At the end, this is just a short summary of the *Industry 4.0* technologies, which can be applied to a *CAD* System (as shown in *figure 2*), included in, or as an integrated surrounding solution, or as information producers for the evolving design process (Muñoz & Perez, 2017).

## 3. DIFFERENCES BETWEEN SURFACE SHIPS AND SUBMARINES

Differences between surface ships and submarines, restricted to the use of CAD Systems, can be grouped in four types: differences consequence of the product itself (the vessel), differences due to the design processes, differences derived from the construction processes, and differences consequences of the boat exploitation (lifecycle).

### 3.1 DIFFERENCE IN THE PRODUCT

#### 3.1 (a) Hull forms

Although submarines´ hull forms are not the most complex compared with other marine artifacts, the existence of bridge fin, skegs, appendages…, see *figure 3*, make submarines more complex than standard vessels from the hull forms point of view.

Figure 3: Hull forms of an attack submarine

The possibility in CAD Systems for importing surfaces defined with third party tools (mainly mechanical systems), valid for naval architecture calculations, could be not enough, as the importing procedures normally imply reformulation of surfaces that make them heavier than in the original format and therefore producing worse performance and further problems in production information (i.e. shell plates development). It means that it is preferable that the CAD System has its own advanced hull forms definition tools, and if it has not them, at least to have optimized links with such those applications.

#### 3.1 (b) Complexity of the vessel

Submarines are so much complex compared with a commercial vessel of similar size, that means much more data. At the same time submarines have more dense spaces, obliging therefore the CAD System to handle huge amount of information. It means that the data structure and the visualization options of the CAD System should be optimized to guarantee a good performance. But the aspect of the complexity of the vessel does not refer only to amount of information, but to the level of detail with which it should be generated in the 3D Product Model. All elements of the 3D Product Model are defined and need to be often times displayed and handle with a very high level of detail, eventually producing therefore more performance problems. This requirement is even more important in the case of submarines, where due to lack of space the 3D Product Model is built very accurate.

#### 3.1 (c) Sister vessels

The concept of sister vessels is not exclusive of submarines. In fact, it is more extended in surface ships, but submarines have some particularities that make more difficult the management of series. The long duration of submarines´ programs and actual construction of units produce that many of the equipment, especially weapons and communications, of first units change in subsequent ones, making thus advisable the use of PLM systems, and obliging CAD System to interface them and to allow sharing date.

#### 3.1 (d) Special rules

Although in the last years there have been attempts by regulatory bodies, classification societies and maritime organisations to oblige the different states to apply international rules and class notation to boats, however usually submarines do not comply with international regulations or rules applicable to surface ships. They are conceived, in most of the cases, following particular rules (intact and damage stability, collision or impact) established by each Navy. Although these rules are in most of the cases stricter and more restrictive than those for surface ships, their particularities make them generally not covered by CAD Systems. This implies the use of specific calculation tools, different from the CAD System used for the rest of the project. In order to avoid this, the CAD System should be enough flexible to incorporate new functionalities, and if this is not possible, at least allowing to interface suitable tools in an easy way and maintaining the integrity and consistency of data.

#### 3.1 (e) Shared projects

The complexity of submarines, as well as the huge investment that they involve lead often to situations in which submarine industry is developed as a result of the collaboration of different companies, not only design offices, but also the manufacturing shipyards. Even more, in latest times is being more and more common to have submarine programs shared not only by different companies, but also by different countries. There are several possibilities for this collaboration (for example splitting of the submarine by zones, by technological concepts, by functions), but all of them contemplate that CAD Systems used by each collaborating part should be the same, or in case this is not possible, at least that both systems could understand each other and interchange information in an efficient and fast way. This also implies a rational handling of legacy data.

### 3.2 DIFFERENCE IN THE DESIGN PROCESS

3.2 (a) Segregated design phases

In surface ships the time is an essential driver, as one of the key factors for a quick manufacturing of the ship. But in submarines, the long duration of the construction program makes the time no critical important. It means more segregated design phases, and in most of the cases minimisation of the use of preliminary or not approved information, thus avoiding the necessity of redoing modelling tasks. Also, it means that drawings and outputs in submarines are not generated according to the just-on-time policy, but in advance to the fabrication phase. This aspect significantly facilitates the use of a CAD System.

3.2 (b) Rigid design process

The nature of the product, its purpose, the severe operation conditions and the material used, as example, oblige to a bigger control of the production process, and the design one. This control implies, among others, the validation of previously defined data, the access control to the 3D model and the record of all interaction with the Database for further tracking (change control). All these aspects should be covered by the CAD System, which should provide innovative solutions without losing any performance or handicapping the efficient work of the designers. Special attention should be put in the maturity control of the design through the validation of the 3D model, as the most suitable tool to measure the degree of advance of the design.

3.2 (c) Moulded line policy

The moulded line policy is one of the aspects in which the design often differs from submarines to surface ships. While in ships, profiles are usually positioned in the moulded line side. In submarines, see *figure 4*, the shell plating thickness is looking inside, so the tapering rules should allow to adjust the profile shape to plate thicknesses when these have different values.

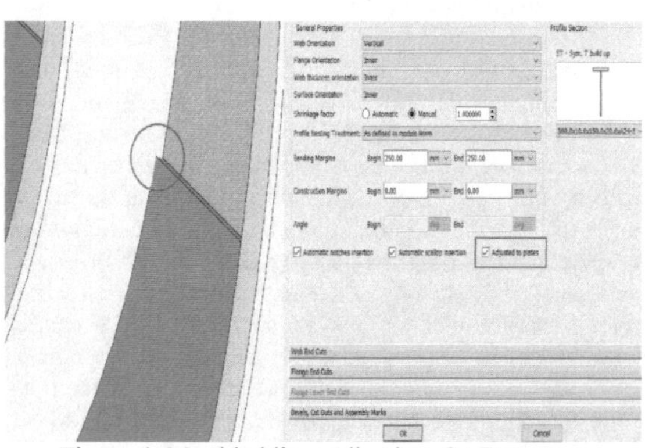

Figure 4: Moulded line policy in a CAD system

3.2 (d) Complex workflow

Submarines are complex platforms in which many participants are involved. Even in design phases it is very common that the design itself is shared by different specialists, each one devoted to a particular task but at the same time handling the boat as a single entity. It is preferable that all the participants would be physically located in a single place, but this is not the most common case. Thus, the CAD System should provide a suitable solution not only for integrating specialists remotely located, but also for integrating the tools used by these specialists. This integration will allow to interchange data, both geometric and technological. The problem exists both in submarines and in surface ships, but in the case of boats the bigger number of participants, tasks and controls make the process more complex.

3.2 (e) Security restrictions

The participation of different suppliers, design agents, work procedures and design tools should not affect the consistency of the information and its confidentiality. Therefore, all data transference and remote access must comply with specific requirements for secure work that produce a reduction in the efficiency of the design. Regarding the CAD System, these requirements mean an additional obstacle as they make more difficult the technical assistance and consultancy services.

A CAD System should have different possibilities as regards the visualization, as certain objects would be shown to certain users by means of a "secure" representation. The aim of this level is to contain the geometry to be shown to non-authorized users during the different design tasks (equipment reading, etc.).

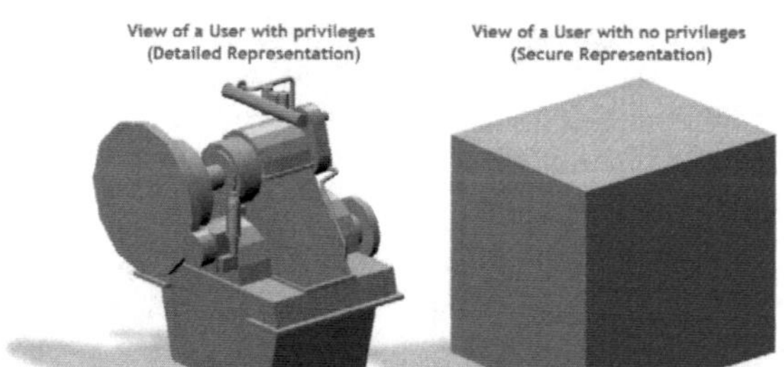


Figure 5: Representation of a library model according with its security level

The Shared, Detailed & Simplified or Secure representation group should be read depending on the user privileges for the library model, as it appears in *figure 5*.

3.2 (f) Analysis of alternatives. Prototyping

The unique characteristics of submarines, the long duration of the design and the necessity to improve existing solutions could derive in the parallel study of different design alternatives. This affects the concept of single product model, and the CAD System shall be able to manage the subsequent situation and to merge approved alternatives. In some cases, different alternatives are consolidated, being also the source for having different projects to be managed simultaneously. The study of alternatives could also imply the use of simulation tools that shall be available in the CAD System, or at least compatible with it.

3.2 (g) Rule based design

As in surface ships the rule-based design is focused mainly in production aspects, trying to optimise fabrication processes, in submarine´s rules are mainly devoted to actual design aspects. CAD System shall allow designers to propose solutions acceptable from that point of view and avoid others contrary to it. And as these rules are very particular for each builder and Navy, CAD System shall be enough flexible in order to be customised and provide suitable solution for each organisation.

3.2 (h) Legacy data

Surface ships implies the design of many prototypes, from which only some of them will arrive to a later stage of the design, and only very few become projects to be actually built. At the same time, there are data from actual projects valid for the future.

All of this means that shipyards and design offices have available data that could be reused for new designs, saving therefore time and effort on them. If the existing data is available in the same format in which the new design is being developed, the data transference is not difficult, as all CAD Systems have export and import tools for own data. But if the data is in different format (information from equipment suppliers, standards,) the process requires not only the data exchange, but also the format translation, direct or through a neutral format. In such a case, it should be taken in consideration the size of the data once imported, as normally it produces degradation in the performance of the CAD System.

3.2 (i) CAD System – PLM system integration

The integration of the CAD System with the PLM system in submarines is needed to perform an integral management of the design, but not only this, also of the product through the whole vessel lifecycle. All departments involved in the design and manufacturing of submarines need to address important requirements like: document management for all related CAD documents and files, maintenance of the relation between the 3D Product Model and the documents generated by the CAD System, management of standards, workflows and processes associated to the design and manufacturing processes, common access control, coordination of project participants, supply of updated and correct information to all parties. It is therefore necessary to make available the CAD information to all departments of the shipyard (non design departments), as operations, planning, production, purchasing, dockyard management... and this is something to be done by linking both the CAD System and the PLM system.

3.3 DIFFERENCE IN THE PRODUCTION PROCESS

3.3 (a) Standardisation

Submarines use particular standards than in many cases are not used in surface ships. This applies to structural materials (steel, alloys or profile cut-ends), outfitting components (pipes, fittings, duct sections, hangers and penetrations), electrical elements (transits and connectors for example), fabrication methods (plate bending, welding, and pipe bending) and assembly units. Standards provided by commercial off-the-shelf CAD Systems are not applicable, and the CAD System shall allow the creation and edition of particular standards. Moreover, it shall provide suitable tools to adapt existing standards to new ones according to shipyard internal procedures and constraints.

3.3 (b) Welds management

Riveting remained the principal method of plate joining until the early part of *World War II* (WWII). At the start of the WWII, welding was being used to join the circumferential butts of pressure hull plates in submarines but seams still were riveted.

Current welding practice for submarines´ manufacture involves a lot of resources, increase the weight and the cost of the boat, so it is necessary to be able to manage efficiently hull structure welding and pipe joins in the CAD System.

In the following image, *figure 6*, appears how easy could be manage the different structure welding contacts that are based on an accurate calculation of geometrical contacts.

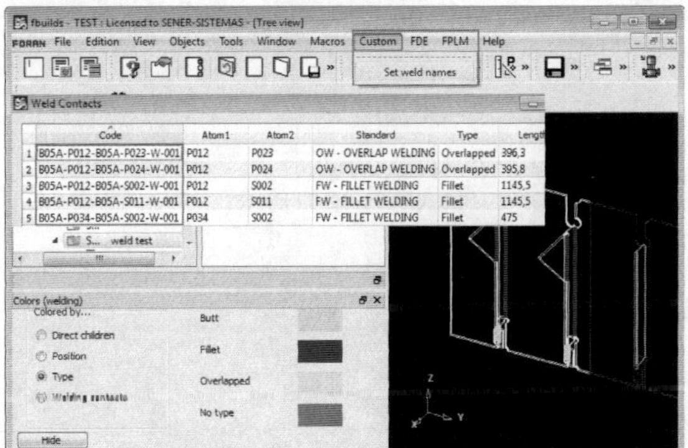


Figure 6: Structural welds management in a CAD tool

3.4 (c) Installation phase (equipment assemblies)
There is an aspect very important for submarines, which consists on handling of groups items in relation with the installation phase. The design in narrow spaces often

require special consideration, and more demanding requirements of the installation sequence (i.e. valve and actuator, fire extinguisher and its associated support, etc.). There are many generic cases where this aspect must be addressed in the design stage.

The first case is related to the auxiliary structures' hierarchy, the modelling process and the way parts can be organized for planning purposes, e.g. ladders with their mounts. Modelling an auxiliary structure in a CAD System gives the possibility to create a hierarchical structure, made of single parts (plates, profiles, special) and substructures (groups of parts). However, planning processes require to being able to organize these parts in a different way that they are modelled.

The second case deals with equipment composed of multiple items, e.g. fire extinguisher and the bracket, or a door and the frame. Normally, in submarines, the equipment is composed by individual items that can be grouped in an interim product.

3.3 (d) Production documentation

The complexity of the design process in submarines have its parallel in the production phase. Fabrication and assembly documentation shall be adapted not only to own requirements of each shipyard or subcontractor, but also to particular needs of submarines, both in the content of the documentation and in the format used. This means that the CAD System, see *figure 7*, must allow to generate outputs (including NC) according to these needs, and this implies that it shall include tools for customising outputs in an easy way.

Figure 7: 2D & 3D visualization of a submarine transversal section

This customisation is also needed because special materials, standards and processes involved imply the use of a specific production machines, so the data should be generated in a particular way to feed that equipment.

Moreover, once generated, in submarines fabrication and assembly documentation follows more rigid flow and more demanding procedures of validation and revision, being almost compulsory a complete *Documentation Management* (DM) System.

Although this DM system usually is not part of the CAD System, at least shall be interfaced with it, so the status of version of each document could be controlled. A strong and efficient method to find associations between items and outputs is an important requirement.

3.3 (e) Production timing

The long duration of projects in submarine industry provides designers with enough time for generating all production documentation.

However, the complex and rigid approval, procurement and testing processes in submarines operate in the opposite direction, obliging to issue information, material take-offs and drawings earlier than theoretically needed as in surface ships.

3.3 (f) Use of different tools

Design is only a part of the building process. There are many other tasks getting the data from the design or providing data to the design that have big influence in the overall building quality, timing and cost.

Although this is common to surface ships and to submarines, in fact the complexity of submarines makes this aspect more important. It is therefore required that the systems used in all tasks shall understand each other, or at least, be able to transfer information in an acceptable way.

Regarding CAD System, it shall provide suitable interface with material management system, planning system and simulation tools.

The paradigm of this problem appears in case there are two or more companies (shipyards or design offices) collaborating in the same project, even at fabrication stage, and each one has own tools. In this case the systems shall provide not only data exchange between CAD System and other systems, but also between different CAD Systems (heterogeneous design in context) leading to different degrees of integration like visualization, spatial integration and cross manufacturing.

3.4 DIFFERENCE IN THE EXPLOITATION OF THE PRODUCT

3.4 (a) Exhaustive monitoring during operational life

In order to comply with the objective assigned, submarines shall remain in optimal operational conditions during all their life and, shall be updated to technological innovations during that time. This implies that modernisation works in systems, equipment, weapons and telecommunications, in bigger or smaller scope, shall be carried out periodically.

Although the control of changes, updates and modernisation are usually controlled and managed with the help of a PLM system, it has also consequences in the CAD System, that mater the definition of most of the 3D model data.

From this CAD System point of view, the as-build 3D Product Model shall not remain frozen, but shall correspond at any moment to the actual state of the vessel in order to be the basis for that modernisation works, so any change in the vessel must be also reflected in the 3D Product Model.

This problem is more complicated as not all built units of a series and corresponding to a particular 3D Product Model are modernised. In that case the 3D Product Model is split and from that moment handled as two different ones. Applicability management is therefore compulsory. The necessity to maintain updated the 3D Product Model of a vessel through all its operational life has an impact in the versioning policy of the CAD supplier, as a version of the CAD System that correspond to the 3D Product Model must be available at any time during this period.

3.4 (b) Simulation tools

The test for checking the validity of any boat design is the response or behaviour of the submarine under particular conditions in its operational life.

For submarines those conditions include some circumstances that are very difficult to test in advance, being thus necessary to simulate them. In order to carry out these simulations, the best option is to get the data from the 3D Product Model, as it corresponds to the actual state of the boat once built.

Although the CAD System may have own tools for performing some simple simulations, more complex simulations should be carried out by special tools for this purpose. Therefore, CAD System shall have tools for exporting 3D Product Model, including metadata, to those simulation tools. Moreover, it would be convenient that CAD System is retrofitted with data coming from the results of the simulations.

During the design phase simulations could affect the design itself, and once the design is finished, manufacturing and progresses, the results would affect the design of future projects or produce an update of the simulated one.

## 4. A CAD APPROACH FOR SUBMARINES

There are many advanced shipbuilding CAD systems in the market, and a few of them are used in several of the most demanding submarine shipbuilders around the world.

Usually CAD tools can be used in both surface ships and submarines, and for any type of ships or boats. It covers all design aspects (forms definition, naval architecture calculations, hull structure, outfitting, electrical and accommodation) and phases (conceptual, basic, classification and production).

Based in the concept of single data base, integration, topology, and concurrent engineering, many CADs provide suitable solutions for particular aspects of submarines.

### 4.1 APPROACH TO SUBMARINES PARTICULARITIES

4.1 (a) Submarines hull forms

CAD systems include several alternatives for definition of hull forms. On top of the traditional methods based in lines, it provides advance solutions based in NURBS surfaces, allowing the generation of any kind of forms, including multihulls, appendages, knuckled forms...

Once the hull forms are defined, it is possible to define decks and bulkheads following any geometry, and perform the naval architecture calculations (hydrostatics, intact and damage stability, floodable lengths or freeboard). At the same time, it is possible to start the creation of the hull structure and outfitting elements.

Any modification in the hull forms consequence of new design or towing tank tests produce an automatic update of the hull structure already defined.

4.1 (b) Complexity of the boat

Topological definition of elements in a CAD database implies that the amount of data to be stored is minimum, allowing a better performance of the system. At the same time, a CAD tool can handle a very complex 3D Product Model without losing performance in a significant way.

Designers have available all the information, existing advanced tools for reading it for reference. On-line interference checking is available, allowing designers to make design consistent from the very beginning, regardless the complexity of the model (see *figure 8*).

Figure 8: A series of cables penetrating a watertight bulkhead, including cable glands, in a submarine

4.1 (c) Applicability

CAD systems require PLM solutions, so they must include tools for interfacing them, thus allowing such PLM systems to manage sister ships (ship applicability) and changes in the configuration. The interfaces should transfer the information from a CAD database that can be read by the PLM system. Also, it should be possible to develop dedicated direct interfaces with particular PLM systems on demand (see *point 3.2 (g)*).

4.1 (d) Submarines´ special rules

CAD systems need to be fully customised. By means of CAD Development Environment, users could create own commands for performing precise tasks and for adapting some commercial off-the-shelf solutions. On top of that, all geometrical and technological data can be exported so third-party software can read and use them for any purpose.

4.1 (e) Shared submarines´ projects

Some CAD tools provide suitable solutions for joint designs from remotely located sites, and for sharing information between design offices. On top of the traditional splitting of work by geographical concepts, there are also other possibilities as database replication (synchronous and asynchronous) and remote database access. A few CAD systems includes also advanced tools for controlling access and changes of the 3D Product Model. In case of joint design carried out by different companies using different systems, data is exported/imported to/from neutral XML format, so each party can visualise the design performed by the other one.

## 4.2 APPROACH TO PARTICULARITIES OF THE SUBMARINES DESIGN PROCESS

4.2 (a) Segregated design phases

The topological definition of the 3D Product Model provides a suitable solution for the propagation of changes coming from an earlier stage of the design. On top of that, the relationship between drawings generated and 3D Product Model guarantees a fast regeneration of drawings affected by changes in the 3D Product Model. Particular tools for detecting affected drawing are also available.

4.2 (b) Moulded line capability

The tapering allows to adjust the profile to plates when the plates have different thickness.

This option is divided into two areas, one referred to profile processing and another one to create tapering rules.

Profile processing inside a generic CAD:

- On thickest plate: the profiles only will adjust to the plate that has the major thickness.
- Adjusted to plates: the profiles will adjust to the plates that are below. In this case the adjustment will follow the tapering rules. If there is not any tapering rule defined, the default representation will be as a step.

Tapering rules in a generic CAD:

- Tapering with step: tapering representation where the thickness difference is not included in the tapering rules.
- Tapering with bevel: tapering representation included in tapering rules.

The tapering rules will affect both profiles and thickness lines. Because of that although the option adjusted to plates is deactivated, the thickness lines will be created with tapering as it is defined in the tapering rules.

4.2 (c) Rigid design process

Some CAD tools provide the possibility to control the access of the users to the 3D Product Model. This access control can affect the whole project, particular zones, specific entities, upon manager decision, and be applied to certain or all users. The locking of a particular zone or system means its freezing for design validation, what is also available. On top of that, there is the possibility to control the changes to be input in the 3D Product Model, and to track all of them by means of user/task combinations. Based on these considerations about access control and locking, third party planning software can handle the data and obtain conclusions about maturity of a project.

4.2 (d) Security markings and Access Control

CAD data are usually stored in a database server and can be encrypted and protected. The necessity to maintain the confidentiality of the design affects mainly to the technical assistance and consultancy works that SENER could provide. In this case, two possibilities are available. Regarding the data itself, it would be convenient to have a replica of the database in a secure and classified area that allows the developers to check eventual problems and to solve them validating the results with actual data. On top of that, it is necessary to have on-site technical staff, with adequate security classification that could solve daily questions derived from the use of a CAD.

CAD access control is composed by the combination of the definition of user roles, which basically determines the tasks that users are allowed to carry out in CAD modules, plus the ability to determine read-write permissions of users or groups of users based in disciplines (types of entities) and according to *activation units*, which depend on the type of entity, but basically consists of a combination of functional and spatial concepts where the items are grouped.

For example, in the case of equipment items, the *activation units* are the combination of the zones and systems where they are defined. This user access rights are later controlled in each CAD design module.

Other advantages in CAD, related with the access control, could be summarized in the following list:

- Easy management of users and groups of users
    - Ability to share the same users for both CAD and PLM tools.
    - Ability to maintain a single user access role across all the boats of the project.
    - Ability to share user access rights in both tools.

- User access should be based on security classification requirements.

- The security classification should be passed across the interface from the PLM system to the CAD.

- As regards the level of granularity, user's access control should be checked at item level.

4.2 (e) Analysis of alternatives. Prototyping

Most of the CADs provide the possibility of defining different views 3D Product Model (see *figure 9*) by means of build strategy trees, being therefore possible to obtain outputs according to them. Apart from this, new projects can be created having as source previous projects, being possible the analysis of alternatives, and once one of them is approved, input the corresponding data in the original database.

Figure 9: Virtual reality operator inside an Engine Room

4.2 (f) Rule based design

A few CAD systems include tools and commands that allow to carry out the design based in some predefined rules.

These predefined rules affect mainly to production processes (manufacturability of pipes and hull structure), and with the help of a CAD Development Environment it is possible also to define actual commands that comply with design rules.

4.2 (g) Legacy data

CAD tools provide export/import tools that allow to transfer 3D data between projects, totally or partially (see *figure 10*). In such a way, users can select the information to be reused in other projects.

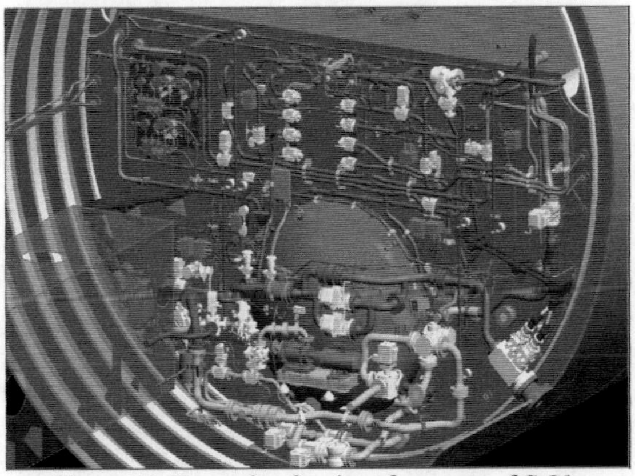

Figure 10: Example of an interface zone of *S-81*. Courtesy of *Navantia*

Special attention should be put on nomenclature and in order to avoid overwriting or inconsistencies. On top of that, CADs can export/import data to/from other software, thus been able of interchanging 3D information with other CAD Systems. DXF, STEP and XML formats are supported.

This 3D information includes all geometric data, and in some cases, also technological data. In order to minimise the impact of imported data in the general performance, there exist the possibility to make from the database references to external files.

As to the migration to CAD latest versions of data created in previous ones, each new CAD version should include tools for updating databases, thus guaranteeing that the databases do not become outdated.

4.2 (h) CAD-PLM integration

A few CADs provide a framework composed of a set of new components allowing to interface CAD with a third-party PLM application.

As example, just to mention the FPLM (FORAN–PLM application), where its components include a Java process (middleware) acting as connection point between the functionality of the FORAN modules and the PLM System, a set of utilities integrated in the FORAN modules to invoke the interaction functions with the PLM, a set of CORBA IDL classes and methods to define the communication between FORAN and the FPLM Server, and one library of FPLM plug-in methods and classes specific for each FPLM.

4.3 APPROACH TO PARTICULARITIES OF THE SUBMARINES PRODUCTION PROCESS

4.3 (a) Standardisation

CADs include tools for definition of any kind of standards. Structural materials (steel, alloys and profile cut-ends), outfitting components (pipes, fittings, duct sections, hangers and penetrations), electrical elements (transits and connectors), fabrication methods (plate bending, welding and pipe bending), assembly units and graphic symbols, can be fully customised and adapted to particular shipyard requirements.

All standards can be stored in kind of master project from which it is possible to extract information to create certain project standards.

4.3 (b) Welds and coupling management

Structure welds functionality in a CAD should cover automatic generation of welds with attributes including the identification of the entities to be welded. Or manual definition of welds in a given point.

An existing weld could be edited to modify the length of this weld along a defined contact. Two or more welds can be joined to create a single weld. And a single weld can be split into multiple smaller welds.

The system allows the definition of welds between objects belonging to different design disciplines (i.e. piping and primary structure). It is possible to assign automatically attributes to welds following user rules. Output drawings and reports can display both visual representation and attribute data.

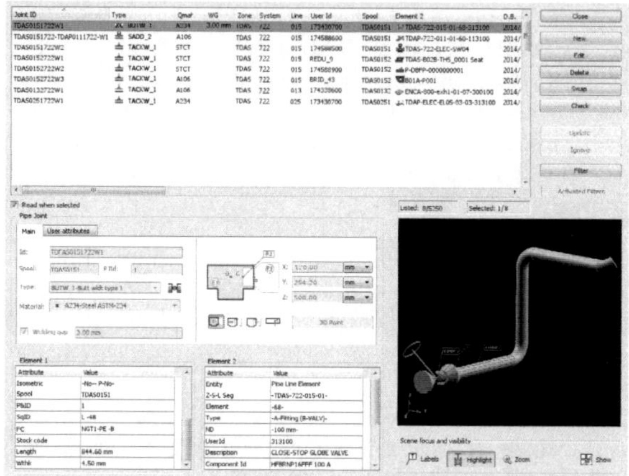

Figure 11: Pipe join manager window in a CAD system

The Pipe Joint, see *figure 11*, will be used to identify the joint between two elements. Although these two elements could be of different disciplines, one of them will compulsorily have to be a piping element.

The second one could be:

- Another piping element.
- A duct line or cable tray line element.
- An equipment item.
- A main structure part.
- A part of an auxiliary structure, or a support.

4.3. (c) Equipment assembly

The equipment assembly is an outfitting entity type added to the CAD. It behaves as a collector of several equipment items, allowing to manage all together as a whole when reading or deactivating from the scene and where each piece of equipment will keep its own component, user identification, properties and user attributes and, therefore, it will be able to behave as a common piece of equipment, in particular, it will be able to be read independently.

The equipment assembly will also have its own component that will be a new type of component in outfitting, with its particular identification and stock code. It will consist of two or more common equipment components where each one of them will have specific position and orientation within the assembly.

For the management of assembly components, there will be a new command, the Assembly Component Manager:

- Assembly components will be organized in a hierarchical tree of libraries where each library will have a unique name in the whole tree. Components will be created under library nodes and they will have unique identifications within their parent libraries.
- Users will be able to manage the assembly component tree; therefore, they will be capable of creating, editing, deleting, cutting and pasting library nodes.
- Users will be able to create new assembly components under library nodes and to delete, cut and paste existing components.

For the management of equipment assemblies as project items, there are commands handling and treatment equipment assemblies through an assembly manager functionality, see *figure 12*.

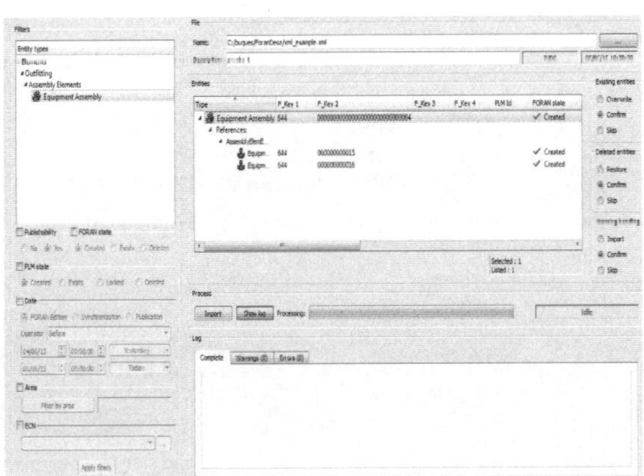

Figure 12: Import/export of equipment assemblies in a CAD system

- The Assembly Manager will be used to create new assemblies, provided that some assembly components have already been defined, and to edit or delete existing assemblies in the project.
- Creating an assembly will mean the creation of an item of this new type of a CAD entity plus an equipment element from each equipment component within the assembly component.
- This manager will also let to split an existing assembly into its constituent items of equipment.
- The Assembly Manager will give users the possibility to position an assembly as a whole or partially, what means that the user will be able to select a subset of its constituents to be positioned.

4.3 (d) Production documentation

All documentation, format and contents, are fully configurable in CADs, depending on the actual needs of each shipyard. CADs may include tools to check the fully correspondence between the 3D Product Model and the drawings generated, and allows to generate NC information to feed fabrication equipment of the shipyard (plate and profile cutting and marking, welding robots, pipe bending...). All outputs can be managed with the help of a DM system, which interface with CAD should be available for customisation.

4.3 (e) Use of different tools

Usually CAD systems are open to export data to third party software once known it. The particular characteristics, features and scope of each system make unfeasible to export all data in general to a neutral format. As the information to be exported is mainly technological, it is necessary to have dedicated interfaces with particular systems.

## 4.4 APPROACH TO PARTICULARITIES OF THE EXPLOITATION OF THE PRODUCT

4.4 (a) Exhaustive monitoring during operational life

At any level of its maturity, including its final stage that corresponds to the operational situation of the boat, the 3D product model can be monitored by means of the own virtual reality module of a CAD. With this, designers and maintenance operators can check all the elements of the submarine, ask for any technological attribute, simulate maintenance operations, apply textures, measure dimensions, simulate special illumination and smoke conditions. This information can be updated, being therefore available all information necessary for monitoring the operational life of the vessel. On top of that, the interface with PLM systems provide a suitable solution for lifecycle management. In such a case, not only all project documentation, but also the 3D Product Model, can be checked and controlled after the design and construction.

4.4 (b) Simulation tools

CAD systems should have the possibility to carry out simple simulations within the 3D Product Model as removal of equipment for maintenance purposes, actual luminance on board or handling of human models. However, it is not a dedicated simulation tool, and there are requirements that are not complied with. For those cases, all geometrical data, together with technological attributes, can be exported to a neutral file (i.e. XML) that can be understood by a real simulation tool.

## 5. CONCLUSIONS

Submarines differ significantly from surface ships, based in the differences that exist in the product itself, the design process, the production process and the exploitation of the product. Nevertheless, the CAD System used for design in surface ships is also suitable to be used in submarines, providing that some adaptation is done. This adaptation means the availability in CAD Systems of solutions to particular boats´ problems, but also some changes in the method the CAD System is used in surface ships.

On the other hand, the improvements and savings in cost and time of surface ships, in part derived from the use of modern CAD Systems, suggest that also submarines´ shipyards and design offices could benefit from the use of such tools, adapting their design and production processes to those ones already in use in surface ships.

## 6. REFERENCES

[1]. BENAYAS, A. & PEREZ, R. (2019). *What should shipbuilding expect from the CAD/CAM systems of the future?*. The Naval Architect. April. pp. 28-31.

[2]. MUÑOZ, J.A. & PEREZ, R. (2017). *CAD tools for designing smart ships in the world of the Internet of Things*. SMART SHIP 2017. London, UK.

[3]. PEREZ, R.; BENAYAS, A. & PEREZ-ARRIBAS, F. (2018). *Data management for smart ship or how to reduce machine learning cost in IoS applications*. SMART SHIP 2018. London, UK.

[4]. PEREZ, R. & PETER, E. (2019). *The use of CAD systems to manage modularity in multi-role warships*. WARSHIP 2019. Bristol, UK.

[5]. PEREZ, R. & GONZALEZ, C. (2015). *History and Evolution of Shipbuilding Oriented CAD Tools*. COMPIT 2015. Ulrichshusen, Germany.

## 7. AUTHORS BIOGRAPHY

**Rodrigo Perez** Ph.D. is the head of *Naval Shipbuilding Business Development* at *SENER Ingeniería y Sistemas S.A.* He is also professor in several subjects at the *Marine Engineering School* of the *Technical University of Madrid.*

# RENEWABLE: AN INNOVATIVE UCL CONCEPT SUBMARINE DESIGN FOR AN ALL-ELECTRIC BOAT

**R Pawling,** UCL Department of Mechanical Engineering, UK
**J Dos Santos, M Yigit Oguz, D Weatherall, B Lango,** UCL SDAC, UK

## SUMMARY

The UCL Submarine Design Course is a three-month short course in which groups of students develop a submarine concept design to meet a broad user requirement. This paper will summarise the eight designs developed by the 2019 cohort, and in particular detail an all-electric submarine for use in the Baltic and North Sea. Codenamed "Renewable", this 1500tonne design features battery-only propulsion with the ability to use ROVs to recharge from a dispersed power infrastructure based around existing and expanding offshore wind farms. This paper will cover the general CONOPS behind the design, the technical solution developed by the students, and conclusions regarding the viability of all-electric submarines in the near future.

## 1. INTRODUCTION

The Naval Architecture and Marine Engineering (NAME) and Marine Research Group (MRG) are divisions of the UCL Department of Mechanical Engineering (https://mecheng.ucl.ac.uk/). NAME is focussed on providing postgraduate and undergraduate education to naval architects and marine engineers, whilst MRG focusses on research topics covering both naval and civil vessels. With a significant overlap in staff, research and teaching interests frequently combine and student design projects commonly feature novel areas of interest. For RENEWABLE these were the increasing energy density of lithium-chemistry batteries, and the potential national security benefits of distributed renewable energy generation.

The concept of an all-electric submarine powered entirely by batteries is not new – the 85 tonne Spanish submarine Peral, commissioned in 1889, could make 3 knots whilst submerged and was armed with a single torpedo tube with three weapons, and was technically successful if not entirely militarily useful [1]. UCL is not the only organisation examining the concept of the modern all-electric submarine, for example the Nevesbu MORAY, which has been developed in conventional, AIP and battery options [2] and the Naval Group SMX 31 / SMX 31E advanced concept [3].

Similarly, at-sea recharging and refuelling of submarines has been carried out in practice, for example the German Type 24 "Milchkuh" (milk cow) submarine of the Second World War. More recently the concept of recharging Uncrewed Underwater Vehicles (UUVs) is being investigated for naval and civil use, and it is notable that the US DoE concludes marine renewable energy is the most viable source for the charging power [4].

## 2. THE UCL SUBMARINE DESIGN COURSE

The UCL Submarine Design Course, also known as the Submarine Design and Acquisition Course (SDAC) is a 9-week postgraduate course, running in the autumn or winter [5]. The first four weeks feature a range of technical lectures from UCL staff and industry experts, covering all aspects of submarine design from stability and structures to weapons and escape. Students work in groups of between two and four to develop a concept design in response to an outline User Requirements Document (URD) provided by the UCL staff. Each requirement has a unique "special feature"; these can be the incorporation of a new technology, an unusual restriction on the design, or a novel concept of operations.

Student groups produce a series of subject-focussed deliverables over the course of the design project, a list of which is shown in Table 1. This practice allows the students to focus on specific technical aspects; assists in staff monitoring of progress; encourages students to "draw a line" under each stage and move on; assists in teaching project management and reflects the use of progress reporting in real-world engineering projects.

Table 1. SDAC deliverables

| |
|---|
| 1. Single Sheet Characteristics |
| 2. Outline Operational Requirements (OOR) |
| 3. Concept of operations |
| 4. System Requirements Document (SRD) |
| 5. Cost Capability Tradeoff |
| 6. Weight and Space Breakdown |
| 7. Sketch General Arrangement (Sketch GA) |
| 8. Firm General Arrangement (Firm GA) |
| 9. Pressure Hull Design |
| 10. Powering Report |
| 11. Stability Report |
| 12. Final General Arrangement (Final GA) |
| 13. Trim and Hydrostatic Report |
| 14. Manoeuvring Report |
| 15. System Line Diagrams (Typically: Electric Power Distribution System, Trim System, Compensating System, HP and LP Air System, BIBS System, Sea Cooling System, Sanitary System, HVAC System) |
| 16. Complementing Plan |
| 17. Cost Estimate |
| 18. Master Data & Assumptions List |
| 19. Risk Register |

The programme includes three design progress reviews and a final presentation to external invitees, and students also have the opportunity to deploy their designs in a wargames day using a UCL-developed tabletop naval wargame [6].

### 2.1 THE 2019 DESIGNS

The 2019 course featured nine fully worked up concept designs, with additional "sketch" designs at a far lower level of detail developed by those students on the lectures-only option. Table 2 summaries the designs, which included conventional, AIP and nuclear submarines, including mini-subs but excluding UUVs (although many designs were to make use of UUVs.

Table 2: Summary of the designs for the 2019 SDAC

| Name | Summary |
|---|---|
| MALLET - A conventionally powered arsenal submarine | An arsenal submarine designed to remain on standby on-station with a large payload of cruise missiles, then conduct a heavy first strike against land targets in support of national strategic goals. |
| NEBULOUS – A Submersible Covert Operations Platform | A nuclear-powered submarine primarily intended to interface with structures on the sea floor, deploying sensor networks, interfering with energy extraction infrastructure or submerged cables, or recovering objects of national strategic interest. Deniability is likely to be a key requirement of these operations. Operations below the submarine's operating depth should be accomplished by deploying large UUVs. |
| OBSERVER – A submarine to counteract hybrid/unconventional warfare | A conventionally powered submarine intended to counter peacetime and hybrid warfare submarine operations as well as prosecute typical ASW/ASuW missions in wartime. Hybrid warfare operations typically rely on deniability; the threat of rapidly making surveillance footage public may serve as a significant deterrent to an adversary. |
| PUNCHY – A submarine to deploy large strategic torpedoes | A submarine to deploy a large, long range torpedo weapon in one of several roles. In an era of increasing missile defence capability, the underwater domain may offer new avenues of attack against heavily defended targets. Such a submarine might complement or replace SLBMs in the strategic deterrence role, allow the attack of very heavily defended naval targets, or allow the covert launching of land attack cruise missiles from unexpected directions or unexpectedly close to the aimpoint. |
| QUICKSILVER – A fast interceptor submarine | A high-speed interceptor submarine intended for short-duration missions, primarily in defence of home waters. By using a high-power-density lead-bismuth reactor plant, the vessel should achieve the maximum possible sustained speed. |
| RENEWABLE – A submarine for a co-ordinated defence | A conventionally powered submarine designed to operate in a group to exercise control of a sea area like the Baltic, defending the homeland from naval activities of a very capable adversary. |
| SHOTGUN – A submarine designed to engage multiple small targets | A conventional submarine designed to fight in congested littoral waters against adversaries who possess a small number of submarines and warships up to frigate size, but make heavy use of small craft, ranging from jetski/speedboats up through FIACs, FACs and corvettes; the smaller craft typically being employed in swarms. |
| TESTBED – a versatile low-cost submarine to support fleet development activities | A minimum-cost submarine designed to support non-warfighting activities, freeing up fleet boats for combat roles. |
| UPSCALE – a single-role, cost-minimised SSN with improved habitability standards | This SSN is expected to undertake the traditional ASuW and ASW missions conducted by SSNs but retention issues require an increased standard of habitability compared to previous classes of submarine. |

## 2. THE RENEWABLE REQUIREMENT

The User Requirements Document (URD) provided to the student group is shown in Table 3. below. To enable a more meaningful cost-capability analysis, the SDAC does not specify a budget per boat; instead, a total budget for the class is specified, along with minimum required numbers. This allows students to trade-off numbers versus individual capability and economies of scale from longer production runs. The concept of "UCL Pounds" (£UCL) is used in the exercise and this paper, as the UCL costing model contains data from a variety of sources including historical designs and modern equipment costs, so is difficult to directly compare against real-world project costs as it does not capture R&D and delay-induced costs that can strongly influence military procurement costs.

Table 3: URD for RENEWABLE

| | |
|---|---|
| RENEWABLE – A submarine for a co-ordinated defence<br>A conventionally-powered submarine designed to operate in a group to exercise control of a sea area like the Baltic, defending the homeland from naval activities of a very capable adversary.<br><br>**1. Primary Role**<br>a. Work with several other boats of the class and shore-based control facilities to prosecute and sink heavily defended surface targets like carrier battle groups or amphibious landing groups.<br>b. Working with other boats of the class, prosecute and sink enemy submarines.<br><br>**2. Secondary Roles**<br>a. Embark, deploy and recover small teams of special forces as required.<br><br>**3. Operational Considerations**<br>a. The operational area hosts a large number of renewable energy generation plants and their associated undersea cabling. This submarine should be capable of recharging its batteries via dedicated underwater charging stations connected to this infrastructure.<br>b. It may be possible for submarines to use the same technology to refuel from tender ships or from each other.<br>c. Several allied neighbouring nations are interested in participating in an integrated program, allowing larger class sizes than might otherwise be typical.<br>d. The submarine should minimise onboard liquid fuel storage, ideally to nil but the use of liquid fuel for emergencies or long transits is allowed is required.<br><br>**4. Constraints / Considerations**<br>a. You have a budget of £4.5Bn for this class. NRE design costs will equal 50% of the first of class build cost.<br>b. Maintain 2+ boats on station in peacetime, surging to 4+ in times of tension. | Capability Summary<br>ASuW – High<br>ASW – High<br>LA – Low<br>SF – Med<br>ISR – Low<br><br>Range – Regional<br>Prop – Conventional, emphasising batteries<br><br>Cost cap: £4.5B for the class (2019 £UCL) |

## 3. CONCEPT OF OPERATIONS

Based on the high-level staff URD, the RENEWABLE student group developed a concept of operations (CONOPS) and Systems Requirements Document (SRD), which translated the URD into solution options and capability levels that could be examined in a cost-capability trade-off.

Table 4: The three levels of conflict and activity for RENEWABLE

| Low level conflict | Medium level conflict | High level conflict |
|---|---|---|
| -Patrol homelands EEZ<br>-Training<br>-Transit in total 100nm<br>-Patrol area 100nm linear<br>-Duration up to 14 days<br>-Underwater recharging | -Sea denial/ blockades/ monitoring<br>-SOF missions<br>-Transit in total 100-600nm<br>-Patrol area 50nm linear<br>-Duration up to 25 days<br>-Underwater recharging | -SOF missions<br>-Seek and destroy the enemy<br>-Transit in total 100-600nm<br>-Patrol area 50nm linear<br>Duration up to 10 days |
| Notable System Requirements | | |
| | Long endurance.<br>Special Operations Forces (SOF) capability. | Minimum of 6 heavyweight torpedoes (HWT) |

The broad CONOPS is based on three levels of conflict with associated activities, summarised in Table 4. Note that the basis assumption is that underwater recharging is not used in the high-level conflict scenario. This does not prevent it from be used, but rather refers to the stored energy and battery sizing.

## 3.1 RECHARGING VIA RENEWABLE ENERGY

RENEWABLE is not the first proposal for a wind-assisted submarine. That honour goes to Robert Fulton's Nautilus of 1800, a 6.48m long submersible driven by a hand-cranked propeller when submerged and a collapsible sail when surfaced. The Baltic sea contains a number of offshore renewable energy installations, particularly wind farms, with plans for further significant investment, e.g. [7]. As well as the obvious potential for technical synergy between an all-electric boat and offshore generation capacity, the concept was seen to have strategic and tactical implications. Unlike centralised thermal power stations, renewables generally make use of a larger number of smaller generators. This makes them more difficult to destroy or disable by a state or non-state adversary. Although it increases the total "attack surface", having more points to defend, each one is only a small part of the generating capability.

It should be noted, however, that offshore generation arrays will generally have a limited number of substations, combining the input of many turbines, and converter platforms, stepping up voltage for transmission ashore, as illustrated in Figure 1. These would of course be adversary targets, but the lower power requirements of a small submarine such as RENEWABLE mean that recharging infrastructure would be dispersed among the turbines themselves, rather than being concentrated at the array-scale.

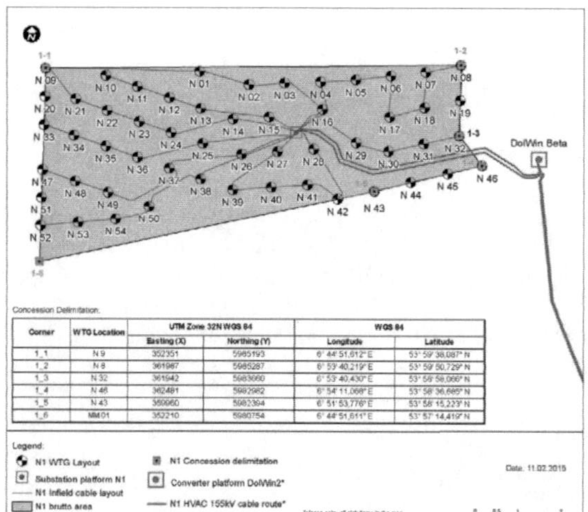


Figure 1: Example layout of a wind farm [8]

Having the ability to recharge at sea would have a number of potential benefits such as extension of mission duration or providing an increased number of sprints per mission. Against an adversary with technical intelligence capabilities, it could also increase uncertainty as to the readiness of boats at sea and prevent them attempting to "wait out" a surge deployment of the submarines.

A specific military downside to using unattended infrastructure is the potential for sabotage. This is potentially more serious than outright destruction as it could include lethal traps for friendly forces. With "dual purpose" infrastructure such as power generation there is a counter in that they can be expected to be regularly checked as part of their usual maintenance. However, this did imply that RENEWABLE needed a way to examine a charging station before approach.

## 3.2 COST BENEFIT ANALYSIS

A cost-benefit analysis was carried out using the UCL COBALT tool. This is a macro-enabled Excel workbook implementing a simple Multi Attribute Value Theory (MAVT) analysis. Students use an Excel initial sixing model to calculate cost increments for different capability changes and COBALT combines these with evaluations of capability increments to generate an overall cost-capability space. COBALT supports up to six benefit areas and six capability areas, with five levels per capability. Incremental costs are calculated using an initial sizing model containing scaling data from the UCL proprietary database and third-party technical document where available. Relative performance of different capability increments against each benefit area can be derived from operational modelling or designer judgement. A simplified MAVT implementation in an Excel tool was developed by UCL to encourage reflective and iterative use of the analysis approach.

In the case of RENEWABLE, the benefit areas were defined as "detection and surveillance", "engagement" and "positioning", with the capability areas being "sonar", "torpedoes", "diving depth" and "battery capacity". The resulting cost-capability space and selected option are illustrated for RENEWABLE in Figure 2.

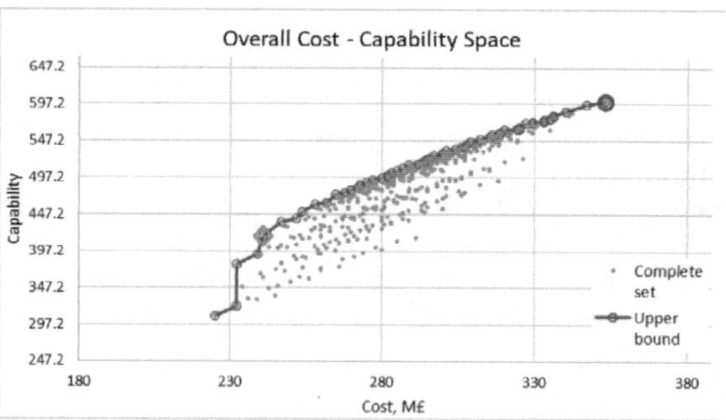


Figure 2: COBALT output for cost-benefit analysis, showing selected option (green diamond)

 

Table 5: Selected options within the capability matrix

| **Level ►** | **1** | **2** | **3** | **4** | **5** |
|---|---|---|---|---|---|
| **Sonar** | small | **med** | med - high | large | |
| **Torpedoes** | **low** | med - low | med | high | |
| **Depth** | 100 | **150** | 200 | 250 | 300 |
| **Battery capacity** | **1** | 2 | 3 | 4 | 5 |

The selected options within the capability matrix are highlighted in Table 5. It is interesting to note that the upper bound of the cost-capability space is made up of a large number of closely spaced points. Although the selected point is at the so-called "knee in the curve", there were no major jumps in cost caused by any single capability level.

## 4. DESIGN SUMMARY

The resulting design for RENEWABLE is a 1493 tonne, partially double-hull two-deck submarine with the principal particulars shown in Table 6. A profile view and main deck arrangement drawing are included in Appendix A.

An overall view of the arrangement is shown in Figure 3, with the casing and pressure hull rendered transparent. The layout is dominated by the four battery banks which occupy the lower deck. Tanks needed for hotel systems and compensation are below the battery deck and all accommodation and operational spaces are on the main deck.

Table 6: Principal particulars of the final design

| **Principal Characteristics** | | **Performance Characteristics** | |
|---|---|---|---|
| Displacement submerged | 1493te | Speed (max / patrol) | 15kts/4kts |
| Displacement surfaced | 1358te | Submerged range | 1300nm (normal), 2000nm (emergency) |
| Length | 60m | Density range | 1.05te/m$^3$ to 1.03te/m$^3$ |
| Beam (hull) | 6m | Stores endurance | 25days + emergency |
| Beam (planes) | 7.75m | Deep diving depth | 150m |
| Operational crew | 15 | Energy storage capacity | 24.3MWh (normal), 37.9MWh (emergency) |
| Accommodation | 15 | Underwater charging ops | Underwater infrastructure or buddy to buddy |
| UPC | 284.2M£ | Underwater charging outfit | 1 x UUV, 1 x cable reel, 2 x docking station |
| Number in class | 16 | Propulsion motor | 1.1MW RDP |
| In service date | 2030 | Manoeuvring thrusters | 2 x 50kW RDP |
| Service life | 32.5 years | Class TLC | 10.34£B |
| **Weapons and Sensors** | | | |
| Number of torpedo tubes | 8 21-inch (533mm) swim-out tubes | | |
| Typical weapons outfit | 7 x heavyweight swim-out, 4 x IDAS missiles | | |
| Masts | ESM, radar, external comms, optronics periscope | | |
| Sonars | 2.5m diameter bow array, medium flank array, mine avoidance sonar, intercept | | |
| Countermeasures | Passive decoy and torpedo hard kill | | |
| Special forces | 2x SOD units, combined escape and SOF LILO chamber | | |
| Underwater comms | Connection available when performing underwater charging | | |

Figure 3: Overview of RENEWABLE with the casing and pressure hull rendered transparent

With only a single deck for accommodation and operations, it was important to develop the design in 3D to account for limited deckhead height and Figure 4 shows the CAD model used to demonstrate the practicality of the arrangement, including representative crew figures.

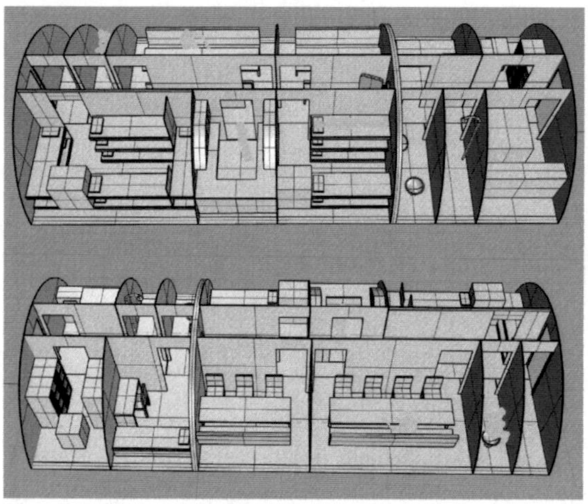

Figure 4: 3D model of accommodation area from starboard (top) and port (bottom)

With two charging stations fore and aft, the associated power electronics is similarly divided, leading to two machinery spaces, one at either end of the main deck. Although RENEWABLE uses non hull penetrating masts, the ops room is still located under the bridge fin. Although the eight torpedo tubes are non-reloadable from inside the submarine, the powering analysis led to a length to breadth ratio that requires their "breech" ends to project into the pressure hull.

Just visible in Figure 3 is the use of a faceted casing; with a substantial casing and flank array in comparison to the diameter of the pressure hull, it was proposed that the external shape be faceted to reduce active sonar returns. The low and wide bridge fin was required to keep within a 10m overall height limitation for shallow water SOF operations whilst still providing enough space to accommodate masts.

## 4.1 WEIGHT, SPACE AND COST

Figure 5 illustrates the weight, volume and cost breakdowns as a percentage of the totals. Note that "variable" includes main ballast tank water. The main impact of the all-battery propulsion is seen in Group 2 – propulsion, which represents just under a third of the boat in terms of mass, space and cost. In contrast Group 3 – electrical is tiny, as no generators are fitted. RENEWABLE has a reserve of buoyancy of 9%

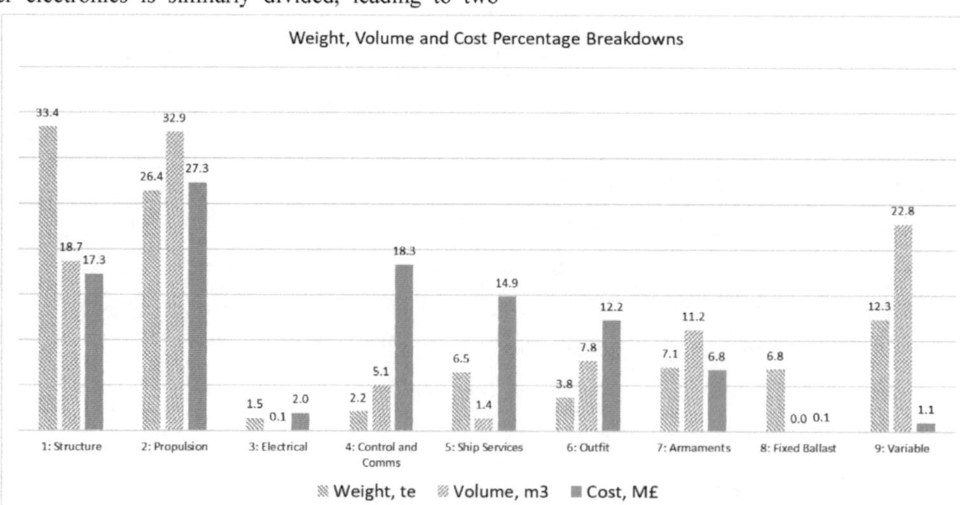


Figure 5: Summary of weight, volume and cost breakdowns

## 5. PROPULSION AND MANOEUVRING

Resistance was calculated using three regression methods, with close alignment between two leading to values of 1.08MW for 15 knots and 75kW for 6 knots. These powers fall within existing sizes of Rim Driven Propulsors (RDP) and so the Voith Inline Propulsor range was used as the basis for the 2.3m diameter main propulsor [9]. An RDP was selected to allow the avoidance of a hull penetration, and for reduced noise.

The need to bottom the boat during patrol led to the adoption of X-planes at the stern, also providing additional redundancy. The resulting plane configuration has an overall span of 7.75m. Depth excursion on a plan jam was calculated as 45m. This is problematic considering the average depth of the Baltic is around 55m. Comparison with existing Baltic submarines (e.g. Swedish Gotland class) indicate the planes are currently oversized and this is confirmed by high or excessive manoeuvring effectiveness indices, indicating potential for refinement.

Although the retractable bow planes were relatively conventional, mounted on the casing, the need for precision manoeuvring and hovering during recharging operations led to the inclusion of two retractable 50kW RDPs on the lower hull. These small propulsors would also be used at patrol speeds of 6 knots, to avoid running the main propulsor at an inefficient low-load.

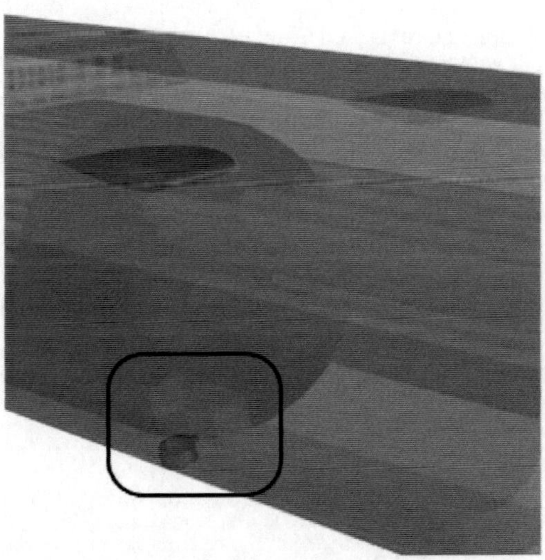

Figure 6: Retractable RDP auxiliary thruster showing withdrawal space

### 5.1 BATTERY CHOICE AND SIZING

With the requirement for a conventional boat (without a nuclear reactor) with no liquid fuels, the two available choices were a radioisotope thermoelectric generator or batteries. The former was ruled out due to the high cost and low power output. Battery performance varies significantly depending on the cell chemistry, as illustrated by Figure 7, which compares specific energy for several battery types.

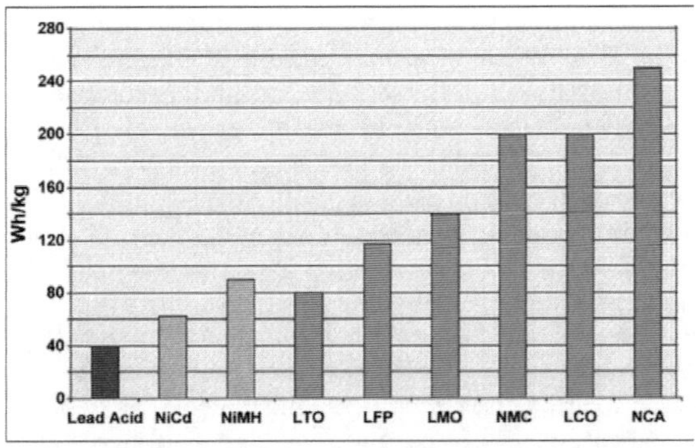


Figure 7: Battery Chemistry Cell Specific Energy [10]

Energy capacity is not the only consideration, however. Buchmann [10] illustrates this using radar plots comparing chemistries over the parameters of; cost; specific energy; specific power; safety; performance; and life span. For the submarine application, safety was the major trade-off with specific energy, so Lithium Nickel Manganese Cobalt Oxide (NMC) was chosen in preference to the higher energy density NCA as the latter has increased risk of thermal runaway [10]. Lithium chemistry batteries are now seeing operational use on submarines, such as the JMSDF Sōryū-class boat Ōryū and are planned for the Italian U212NFS, although the exact battery chemistry for those is unknown.

A representative NMC battery bank was sized using data on Corvus Energy Blue Whale installations [11], as this includes supporting equipment and so has lower specific energy values than those for the basic cells shown in Figure 7 above. The operational profile and assumptions used in battery sizing, along with the resulting size, are summarised in Table 7 below. Unsurprisingly for a submarine with a slow patrol speed, the hotel load, (estimated at 93kW total connected, around 40kW typical), was a similar magnitude to the propulsion load.

Table 7: Battery sizing parameters (High level conflict scenario)

| | |
|---|---|
| **Discharge limit** | 80% |
| **Energy reserve** | 20% |
| **Transit phase** | 300nm @ 6 knots (50 hrs) |
| **Loiter phase** | Bottomed for 10 days |
| **Stored sprints** | 1 x 1 hour at 15 knots |
| **Transit phase** | 300nm @ 6 knots (50 hrs) |
| **Energy stored** | 38MWh |
| **System weight** | 361 tonnes |
| **System volume** | 330 $m^3$ |

## 5.2 BATTERY ARRANGEMENT

Advanced lithium-chemistry energy storage systems such as the Blue Whale system selected as the reference for RENEWABLE require substantial supporting equipment, protective enclosures to prevent the spread of fire and access for maintenance. As such they are not amenable to the historical practice of fitting many small lead-acid cells to confirm with the curvature of the hull. Figures 8 and 9 show the arrangement adopted. The battery was split into four banks longitudinally and three units in each bank; port, starboard and centreline. Although this provided for large units, the enclosures are still bespoke to RENEWABLE. However, given the large production run of 16 boats the cost implications of this should be reduced on a per-boat basis.

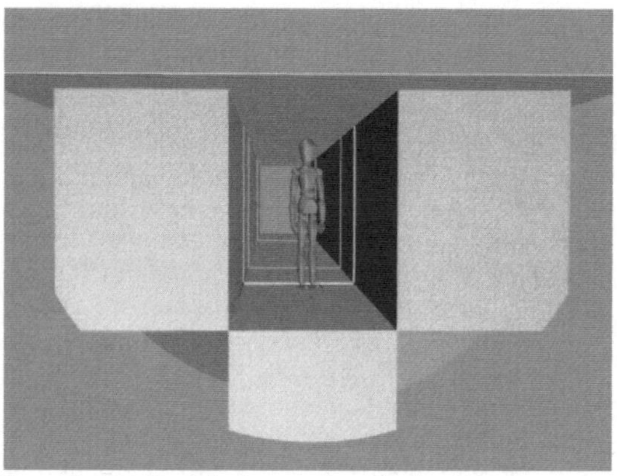

Figure 8: Detail of the battery installation showing access

A further design impact of the need to accommodate large modules rather than small cells is a *consequential* loss of usable volume – Figure 9 shows how the various liquid tanks had to be located in the corners of the lower hull under the batteries, although this is not entirely unusual, being similar to the Type 209.

Each of the four battery banks is in a gastight compartment and provided with a water mist fire-fighting system (in addition to any systems within the modules). Water mist has been noted to be preferable for suppressing lithium-ion battery fires [12] and this system could potentially be combined with additives to increase effectiveness, but remains an area of technical risk.

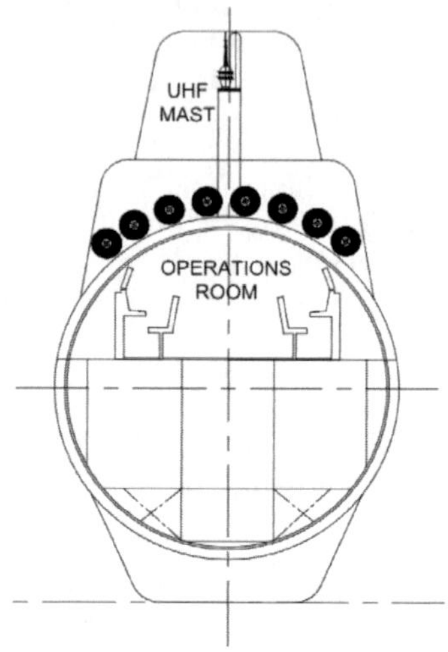


Figure 9: Section in way of operations room, looking aft, showing batteries occupying lower hull with tanks in bilges.

## 5.3 ELECTRICAL SYSTEM

With direct current (DC) battery output, it was decided to use a DC distribution system operating at 1.1 kV. A ring distribution system is divided into a forward and aft section, divided at the single pressure tight bulkhead. A forward and aft bus tie can divide the system into port and starboard busses. Two switchboards are situated in both the forward and aft sections of the submarine. The system line diagram is shown in Figure 10 below, including connections for underwater charging forward and aft and a single connection for shore power when alongside.

 

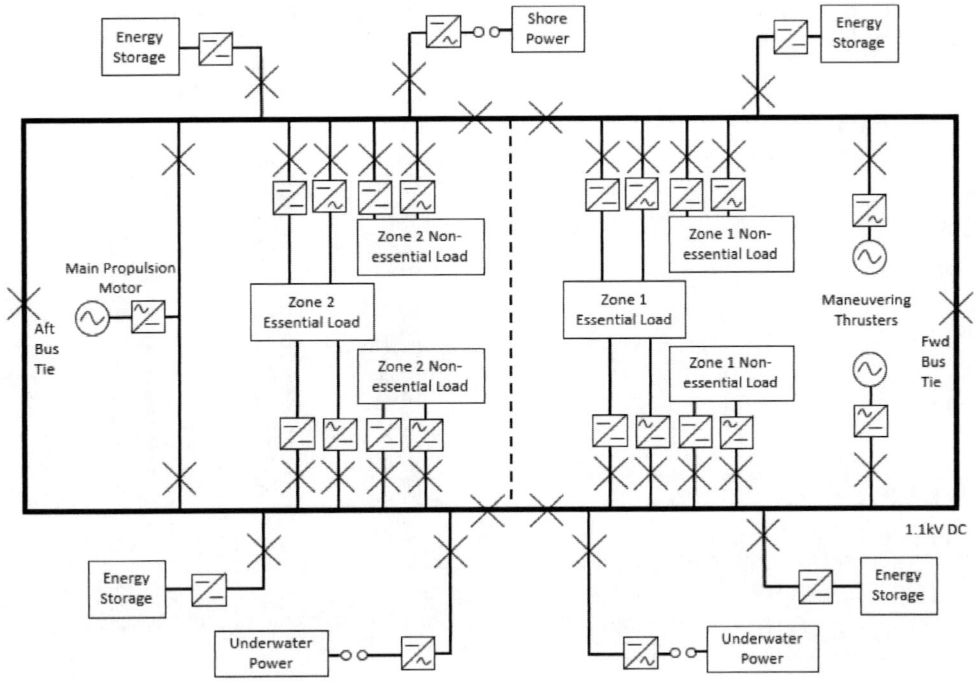


Figure 10: Electrical system line diagram

## 6. OFFBOARD RECHARGING

### 6.1 SYSTEM DESIGN

A key feature of RENEWABLE is the concept of recharging at sea and this has both technical and operational considerations. A wide range of underwater charging methodologies were considered including:

Torpedo Tube Swim out UUV: this has the advantage of flexibility as it would be able to be used in any of the torpedo tubes. Its operation however, is complicated by the fact that it would need to locate and manipulate charging wires separately, as well as complications with recovering the UUV once charging was completed.

Buoyant Wire Snag: this was considered as a way to make an underwater connection without relying on a UUV. However, further investigation into the various methods to safely perform this operation quickly revealed that any solution with this technology would be highly complex.

Wireless Charging: was considered for its relatively simple method of power transfer without connection, however, this method was discounted due to the high power transfer requirements, issues with sea bed sediment build up, the high infrastructure cost, and impact on the submarine design.

Casing Docked UUV: these have the advantage of allowing the charging cable to be integrated into the docking solution, and could be based on existing or near-term technology, so were adopted as the solution.

The UUV and dock system was based on the Saab Seaeye Sabretooth [13]; a compact UUV and docking system, illustrated in Figure 11. The low profile of this UUV is particularly attractive as it can be accommodated under the casing. RENEWABLE is fitted with one UUV with a cable reel, and two UUV docking stations, the concept being that, with each boat and some or all charging stations having a UUV, there will be sufficient redundancy to allow a connection to be made.

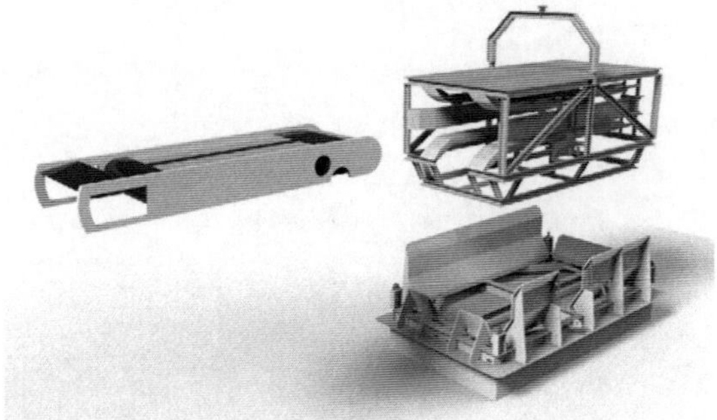

Figure 11: Docking system based on Saab Seaeye Sabretooth [13]

The two-core cable was sized to take a current of 500A, with two connections being used to achieve the required recharge rate. As the neutrally-bouyant UUV could not handle a heavy cable, buoyant sheathing was added to achieve a neutrally buoyant design. This increased the outer diameter of the cable to 75mm, which increased the cable reel diameter. The reel dimensions were calculated based on cable thickness, length and resulting number of wraps, and it was possible to fit it - just - in the casing, as shown in Figure 12.

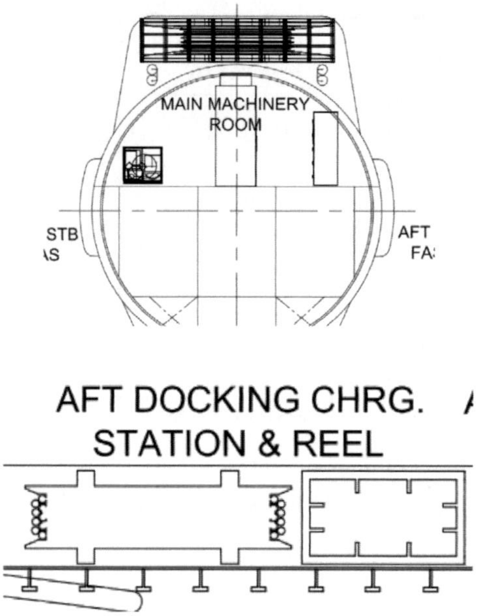

Figure 12: Arrangement of aft docking station and cable reel in casing

## 6.2 CHARGING OPERATIONS

RENEWABLE is intended to operate in one of two at-sea charging operations, summarised in Figure 13. "Buddy-buddy" recharging (A) allows one boat to recharge another. The usual mode of operation would be (B) where the boat uses its UUV to connect to a seabed charging station. Seabed stations could also be equipped with their own UUVs to charge RENEWABLE.

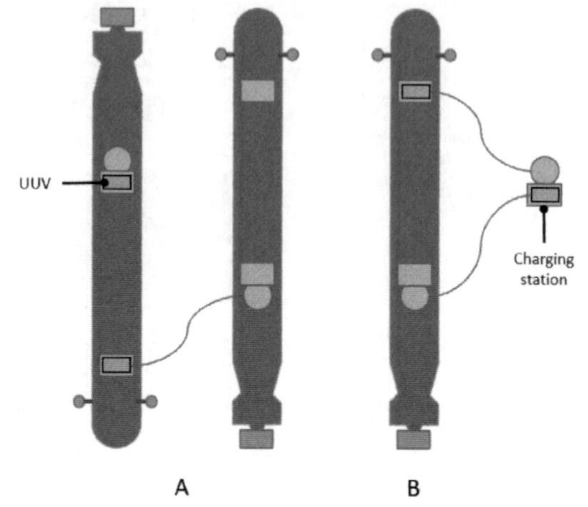

Figure 13: The two modes of charging operation considered for RENEWABLE

The choice use of two connections and associated cable sizing, was driven by the operational modelling. The ability to recharge at sea provides great operational flexibility, but to conduct engineering sizing calculations a scenario must be assumed, and this was the "medium threat" profile illustrated in Figure 14 below.

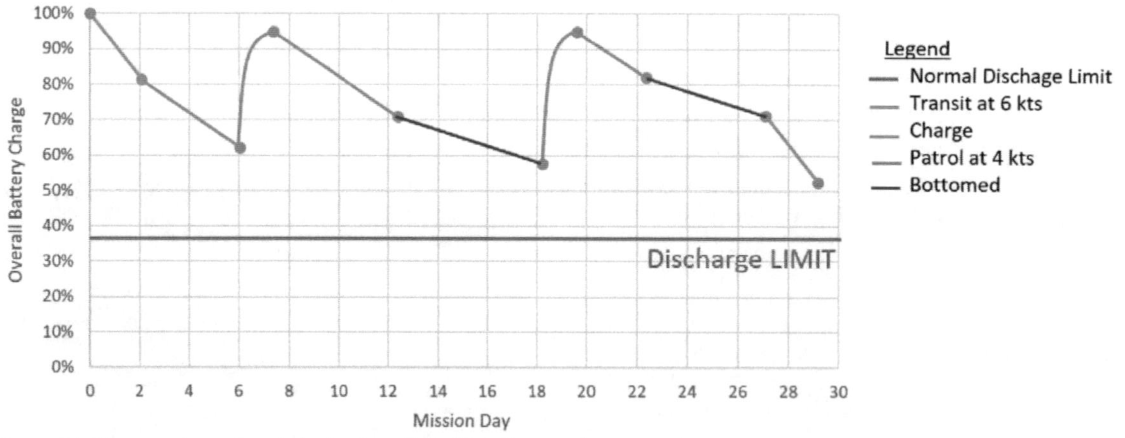

Figure 14: Battery charge levels in the "medium threat" scenario

In this scenario the boat has a maximum of 36 hours to recharge the battery. The electrical system and cables were sized to allow charging in 24 hours, requiring an input power of just over 1MW. This is much lower than the installed power per turbine, which is typically between 3-6MW (e.g. SWT-3.6-120 for Baltic 2, [14]). However it can be seen from Figure 14 that with the significant time the boat spends bottomed, a range of charging strategies could be used, preserving battery energy for repositioning and sprints.

## 7. STRUCTURES

Operating primarily in shallow waters, a deep diving depth of 150m was chosen for pressure hull design. The primary structure is conventional with internal frames and the 44m long pressure hull is divided into two 22m long compartments with torispherical ends. HY-80 steel was chosen, with a minimum safety factor of 1.5 on interframe collapse, increasing to 2.3 on dome collapse. Hull plating thickness was calculated as 19mm with a frame spacing of 560mm. Dome plating thickness was 25mm.

## 8. SYSTEMS DESIGN

Although the hotel systems were broadly similar to a conventional submarine of this size, there were some unique features related to the concept and area of operations.

### 8.1 SANITARY SYSTEM

The operational area of the Baltic Sea and the North Sea, along with frequent bottoming of the boat make production of fresh water challenging. Both due to this and in order to reduce complexity (and hence maintenance) and power demand a "closed" system was adopted where all fresh water is embarked in port. Similarly, grey and black water are not treated and discharged but stored for disembarkation in port. A small amount of seawater is taken aboard during the patrol for toilet flushing. The simplified water system has split tanks fore and aft of the main bulkhead.

### 8.2 HVAC SYSTEM

The reduction of maintenance demands through simplicity and reduction of power demands were key drivers in the design of the HVAC system. Oxygen is stored as liquid oxygen in tanks external to the pressure hull. The additional risk of handling LOX alongside was deemed acceptable as it removed the fire risks from chemical oxygen generators when at sea. Air Handling Units (AHU), providing filter, fan, chiller, heater, $O_2$ dosing and $CO_2$ scrubbing are in each of the two compartments. The decision was taken to permanently split the HVAC system at the main bulkhead to reduce the number of large diameter penetrations. The battery compartments are gastight but connected to the HVAC system to permit entry for maintenance.

## 9. STABILITY AND TRIM

### 9.1 STABILITY

With such a large battery capacity, it was expected that the VCG would be low and stability requirements easy to meet. However, as the design was developed, the limitations on battery shape and requirements for access discussed in section 5.2 both reduced the effective density of the energy storage and increased its Vertical Centre of Gravity (VCG). The final design did meet the requirements of DEFSTAN 08-102 in surfaced and submerged conditions, but only marginally, with the location of the VCG being critical. The results are summarised in Table 7.

Table 7: Summary of stability analysis results.

| DEFSTAN 08-102 Criteria | Threshold | Actual |
|---|---|---|
| Upright GMt | >0.3 m | 0.329 m |
| G> at 30deg | >0.15 m | 0.159 m |
| Area under GZ from 0 to 30 deg | >0.042 mrad | 0.042 mrad |
| Area under GZ from 0 to 40 deg | >0.070 mrad | 0.074 mrad |
| GZ max angle | >50 deg | 97.411 deg |
| **Submerged Critera** | | |
| Submerged BG | >0.3 m | 0.351 m |

### 9.2 TRIM AND COMPENSATION

The lack of liquid fuels and choice of a "closed" fresh and grey/black water system significantly reduced the size of compensation tanks required. However, the seawater density variation over the area of operation (Baltic to North Sea) introduced additional an compensation requirement. In initial sizing, the compensation system was sized to operate in seawater densities of between 1.005 and 1.030 tonnes/m3 (i.e. 0.025tonnes/m3 variation) but, as with stability, the increase in volume required for the battery modules impacted this. Volume for compensation tanks was limited and the resulting density range was 0.015tonnes/m3 with solid ballast being changed for national variants.

The main drivers for the trim and comp system and resulting trim polygon are shown in Table 8 and Figure 15. Note the excessively long trim polygon resulting from the use of long and thin trim tanks in the bilges.

Table 8: Main drivers on trim and comp system

| |
|---|
| 2000 litres of seawater for flushing toilets |
| 1.6 tonnes nett change from expending weapons |
| $3m^3$ of water for SOF LiLo |
| 1 tonne of SOF and equipment |
| 0.015 tonnes/$m^3$ seawater density change |
| 1 tonne minimum freshwater for firefighting |

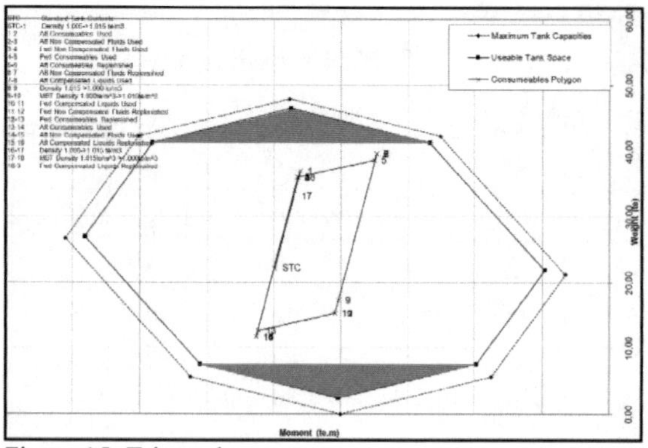

Figure 15: Trim polygon

## 10. DESIGN RISKS

Two key deliverables are the Master Data and Assumptions Risk (MDAL) and Risk Register, as subsequent use and development of an early-stage concept design relies as much on what questions have not been answered as it does on the engineering calculations so far carried out. The three main areas of risk in the RENEWABLE concept are; the all-battery concept; use of NMC lithium batteries; and underwater recharging, each representing different types of risk.

With regards to the all-battery concept the main area of risk is the estimates of propulsion and hotel load – if these are under-estimated then the low energy density of the battery will not just impact indiscretion ratio (as it would for a diesel-electric submarine) but also significantly impact the overall mission endurance. All-battery propulsion also has a financial impact – the propulsion system being 27% of the UPC for RENEWABLE. It should be noted however that the "fuel" costs will be far lower through life, as is being demonstrated with the growing use of electric vehicles [15].

The battery chemistry is considered separately as it has different impacts. Regarding the fire safety aspect, in RENEWABLE this has been addressed by segregation of the battery into gas-tight compartments, provision of a fire-fighting system and using energy densities representative of batteries including enclosures and battery management systems (BMS). The latter is a key component of lithium-ion battery systems in applications ranging from mobile phones to electric vehicles and both ensures safety and improves cell and battery capacity and lifespan. BMS can have many different levels of capability [16] and they would need to be integrated into operator, maintainer and fleet support thinking and practice.

The third area of risk, underwater charging, is a one of practicality and cost. Using an ROV to connect a cable should be well within the body of knowledge for the offshore energy industry. However, doing so from a 1500 tonne submarine in shallow water, subject to currents, in close proximity to either another submarine or a wind turbine pile, may require a new boat handling skillset. There is also the cost implication of how many turbines could be provided with the recharging connections and UUVs.

## 11. CONCLUSIONS

The UCL Submarine Design Course frequently features novel and challenging requirements set by the staff to examine developing and future submarine technologies and concepts. RENEWABLE, developed in the autumn/winter 2019 course, is a conventional submarine with two key novelties; it is entirely battery powered, having no AIP or diesel engines, and being designed for operation in the Baltic and North Sea, it makes use of UUVs for recharging at offshore renewable energy infrastructure.

The concept design development identified several risks but found no fundamental "show-stoppers" to such a submarine, The Lithium Nickel Manganese Cobalt Oxide (NMC) battery system was found to dominate not only the cost and weight, but also the internal arrangement. In contrast to early expectations that batteries would be heavy but not space-hungry, it was learnt that the system and enclosure requirements of modern high energy-density lithium chemistry batteries make them difficult to fit within the limitations of a circular section pressure hull and so the effective volume requirements were much greater.

It was determined that the cables, UUVs and power requirements to support the at-sea recharging concept were physically compatible with a small conventional submarine, indicating that this concept can be scaled beyond previous proposals for at-sea recharging of uncrewed vehicles. Within the context of the design exercise, operational analysis was limited to that needed to size the battery system, but the outputs indicate at-sea recharging could be used to support operational flexibility in several ways.

Finally, it was noted that although the cost-benefit analysis for RENEWABLE led to the selection of a relatively low set of combat system capabilities, the overall solution space appears relatively linear, so scaling the all-electric RENEWABLE concept to larger sizes of submarine should be viable.

## 12. ACKNOWLEDGEMENTS

The downside to teaching is that it is not always possible to stay in contact with the many students who take a course. The first author was only able to reach the second author during the writing of this paper. The other members of group RENEWABLE are included as authors to acknowledge that this paper is ultimately her attempt to retell their work and words.

 

## 13. REFERENCES

1. Wikipedia, (2021), https://en.wikipedia.org/wiki/Spanish_submarine_Peral accessed 2021-05-09

2. Los, S, (2021), "Fully Electric (Battery/Fuel Cell) Powered Submarine", UDT Stockholm 2019

3. Naval Group, (2020), https://www.naval-group.com/en/naval-group-reveals-smx-31-e-its-2020-concept-ship-800, accessed 2021-05-09

4. DoE Office of Energy Efficiency and Renewable Energy, (2019), "Underwater Vehicle Charging", in: Powering the Blue Economy: Exploring Opportunities for Marine Renewable Energy in Maritime Markets, April 2019

5. UCL, (2021), https://mecheng.ucl.ac.uk/study/other/programme/submarine-design/ accessed 2021-05-10

6. Bradbeer, N. & Manley, D. (2017) "The Application of Wargaming to Education in Naval Design & Survivability" Connections UK, 7 Sep 2017, London

7. Wind Europe, (2021), https://windeurope.org/newsroom/significant-developments-on-offshore-wind-in-the-baltic-sea/ accessed 2021-05-09

8. Nordsee One GmbH, (2015), https://www.nordseeone.com/wind-farm/wind-farm-layout.html, accessed 2021-05-09

9. Voith. (2019). Voith Inline Thruster & Voith Inline Propulsor, from http://voith.com/corp-en/drives-transmissions/voith-inline-thruster-propulsor.html Retrieved November 4, 2019

10. Buchmann, (2021), https://batteryuniversity.com/learn/article/types_of_lithium_ion accessed 2021-05-09

11. Corvus Energy, (2019). "Corvus Blue Whale", Retrieved November 4, 2019, from https://corvusenergy.com/products/corvus-blue-whale/

12. Ghiji, M, Novozhilov, V, Moinuddin, K, Joseph, P, Burch, I, Suendermann, B and Gamble, G, (2020), "A Review of Lithium-Ion Battery Fire Suppression", Energies2020,13, 5117; doi:10.3390/en13195117

13. Saab, (2021), https://www.saabseaeye.com/solutions/underwater-vehicles/sabertooth-double-hull accessed 2021-05-09

14. Deutsche Windtechnik, 2021, https://en.wind-turbine-models.com/turbines/669-siemens-swt-3.6-120-offshore accessed 2021-05-09

15. DriveElectric, (2021), "Electric Vehicle (EV) Cost Comparison", https://www.drive-electric.co.uk/ev-electric-car-cost-comparison/ accessed 2021-05-09

16. Hoeger, T, (2021), "BMS Design for Lithium-Ion Batteries, A Holistic Approach", International Battery Seminar and Exhibit, March 9-11 2021, https://www.internationalbatteryseminar.com/battery-management-systems/17 accessed 2021-05-09

## 14. AUTHORS BIOGRAPHY

**Dr. Rachel Pawling** is a Lecturer in Ship Design at UCL, teaching the subject of ship design to undergraduate and postgraduate students. She obtained her PhD in computer aided ship design in 2007 and has subsequently continued her research in projects funded by the EC, UK and US governments.

Team RENEWABLE consisted of:

**João Santos** – Portuguese Navy
**Bjorn Lango** – Norwegian Defence Materiel Agency
**David Weatherall** – Royal Canadian Navy
**Mehmet Yigit Oguz** – Savunma Teknolojileri Mühendislik (STM)

## APPENDIX A: RENEWABLE GENERAL ARRANGEMENT (PARTIAL)

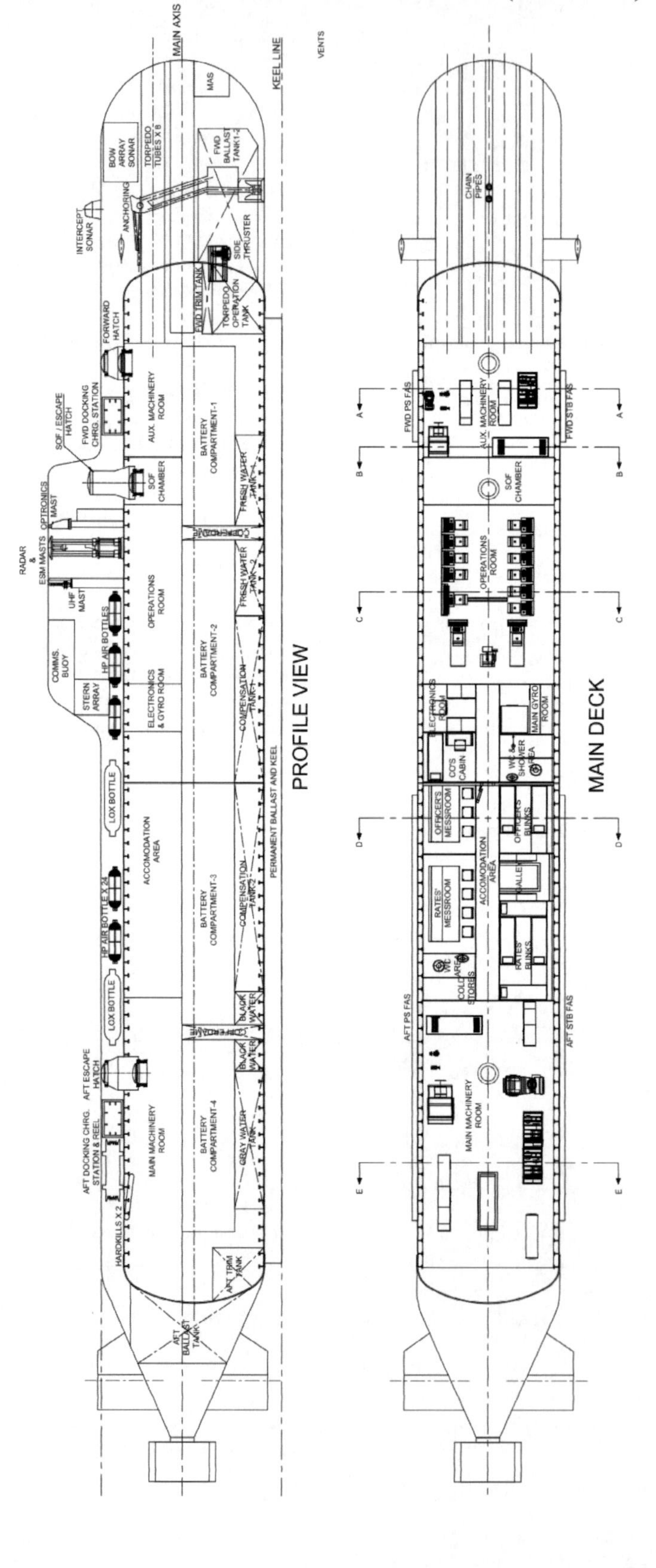


 

# REFERENCE DESIGNS: AN AID FOR ENHANCED ENGAGEMENT IN SUBMARINE PROCUREMENT

**A E Walchester**, BMT, UK

## SUMMARY

Submarines are generally regarded as some of the most complicated and expensive systems in the world. Their procurement is characterised by long production timescales, a vast number of requirements and competing stakeholder priorities. This is compounded as a submarine design is relatively intolerant to change, such as requirements change during the design process or new requirements added to an existing design. These changes can quickly lead to an unbalanced design and a divergent design spiral. Correcting an unbalanced design can prove costly, delaying programme timescales and creating issues that last well into the submarine's in-service life.

The term 'reference design' used in this context refers to a parallel design process that runs alongside the main procurement efforts. It is used to enable the customer to define a balanced requirement set and make informed decisions as part of a collaborative and affordable procurement process. The design is a bespoke pre-concept or concept design that meets the technical requirements of the submarine procurement. BMT explores the benefits of a reference design, focussing on how it can be used as part of the submarine procurement process.

## NOMENCLATURE

| | |
|---|---|
| CADMID | Concept, Assessment, Demonstration, Manufacture, In-Service, and Disposal |
| ConEmp | Concept of Employment |
| MOTS | Military Off-The-Shelf |
| DLOD | Defence Lines of Development |
| TEPID OIL | Training, Equipment, Personnel, Information, Doctrine, Organisation, Infrastructure and Logistics |
| B2TC | Batch Two Trafalgar Class |
| SSK | Conventional diesel-electric patrol submarine |
| AIP | Air Independent Propulsion |
| AUV | Autonomous Underwater Vehicle |

## 1. INTRODUCTION

### 1.1 SUBMARINE DESIGN & PROCUREMENT

Submarines are generally regarded as some of the most complicated and expensive systems in the world. For example, the procurement of the Dreadnought Class Ballistic Missile Submarine "*is the most complex and high value engineering programme underway anywhere within the United Kingdom*" (UK Ministry of Defence, 2015).

Additionally, a submarine design is relatively intolerant to change, such as requirements change during the design process or new requirements added to an existing design. These changes can quickly lead to an 'unbalanced' design and a divergent design spiral. An unbalanced design is one which is hydrostatically unsound or cannot provide enough of a key element of the design, such as space, power or cooling. Correcting an unbalanced design can prove costly, delaying programme timescales and creating issues carry over into the submarine's in-service life.

An example of the impact of correcting an unbalanced design can be found in the procurement of the Spanish S-80 class of submarine (García-Monedero, 2019). A hydrostatic imbalance required the submarine to be lengthened by 10 m, which in turn impacted the infrastructure needed to support the submarine in-service. Overall, the cost of rectifying the problems with the submarine design is estimated to be 14 million euros, with a delay to the program of approximately 7 years. If the decision is taken to extend the existing submarine pens to house the class, then this is estimated to cost an additional 17 million euros (GlobalSecurity.Org, 2020).

### 1.2 REFERENCE DESIGN DEFINITION

The definition of a 'reference design' used by this paper refers to a parallel design process that runs alongside the main procurement efforts. Typically, the design is a bespoke pre-concept or concept design that meets the technical requirements of the submarine procurement. This design is managed by the procuring body and can be used for:

- Requirements' validation;
- Comparison;
- Impact assessments;
- Acceptance planning;
- In-service & disposal planning.

The reference design does not need to be constrained by specific shipyard manufacturing limitations, such as restrictions on the overall submarine dimensions due to the infrastructure. Therefore, the reference design presents the 'art of the possible' rather than 'art of the practicable'. This gives more flexibility in accommodating the requirements and makes it shipyard agnostic.

A reference design is not a digital twin, as it is not a representation of the submarine that will be accepted into service. However, a reference design can act in a similar manner to a digital twin early in the procurement process. At such an early stage, a true digital twin may be unavailable, or it may be beneficial to have a separate digital model to that provided by the manufacturer.

Use of a reference design is not exclusive to submarines, but the design complexity and relative intolerance to change mean that many submarine procurement programmes would benefit from their use.

An alternative definition that is not discussed in this paper is that a 'reference design' describes a design that contains the essential characteristics and is intended to be copied by a third party, but the third party may enhance or modify the design. An example would be a concept submarine design provided to a shipbuilder, where the shipbuilder produces the detailed design and constructs the submarine.

## 2. ISSUES IN SUBMARINE DESIGN & PROCUREMENT

### 2.1 NOTABLE ISSUES

Due to the political interest and the technical complexity, it is unsurprising that there are many examples of issues arising during the procurement of a submarine. Two separate examples which will be referred to in this paper occurred during the procurement of Australia's Collins Class and the UK's Astute Class submarines.

### 2.2 THE COLLINS CLASS

The procurement of the Collins Class submarines comprised of two parallel activities. The first was for the procurement of the submarine, whilst the second was for the procurement of the combat system including the sonar fit (Schank *et al*, 2011a). When the combat system was integrated into the submarine, it was discovered that the large bow sonar array caused noise issues due to poor hydrodynamic flow. This issue was rectified in-service by installing a fairing around the aft part of the array (Yule & Woolner, 2008).

### 2.3 THE ASTUTE CLASS

The design and build of the Astute Class submarines have identified many lessons that can be learned from. One of the decisions that has most shaped the design and in-service life of the class was to expand the diver capability present in the initial design (Schank *et al*, 2011b). The requirements change resulted in a change to the contracted design with an associated impact.

## 3. USING A REFERENCE DESIGN

### 3.1 ACROSS THE CADMID CYCLE

The product lifecycle of defence equipment in the UK is often referred to as the Concept, Assessment, Demonstration, Manufacture, In-Service, and Disposal (CADMID) cycle (Overt Defence, 2020a). Each phase represents part of the lifecycle and should be considered during a procurement (such as for a submarine).

A reference design can be used directly in the initial CADMID phases and can contribute to the planning for the others. As shown in Figure 1, the reference design can be used in requirements generation, design selection and tailoring, and as a planning tool for assessing the impacts and requirements related to in-service and disposal.

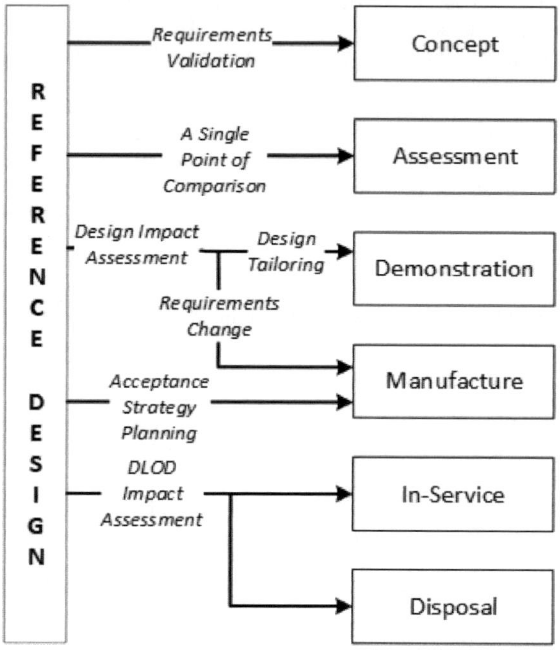


**Figure 1: Reference Design Usage Across the CADMID Cycle**

### 3.2 REQUIREMENTS VALIDATION

There is much to be gained from taking an initial set of requirements and using it to produce a reference design. If the design process produces a design that fulfils all of the requirements, then there is assurance that the requirements are attainable. If a balanced design cannot be produced, then it is clear that the requirements must be revisited. In either scenario, the process will highlight how the requirements interact with each other and where trade-offs need to be made.

This process can be conducted by a shipbuilder producing a prospective design without a clear customer, i.e. where the shipbuilder has to create their own requirements. More commonly, it can be conducted by a procuring body wanting to ensure that the requirements they release are attainable and will give them a submarine that fulfils their Concept of Employment (ConEmp). The ConEmp describes how the desired capabilities provided by the submarine are intended to be employed across a range of scenarios (UK Ministry of Defence, 2011). It is also important for the procurement body to consider how compliance with the requirements will be demonstrated, as covered by the right-hand side of the typical system engineering 'V-diagram' (US Department of Defence, 2001).

Therefore, production of a reference design verifies that the requirement set and the parent ConEmp is balanced, attainable and complete. Lessons learned from producing the design can be fed back into the requirement set to optimise the requirements ready for tendering. In addition, the reference design can be used as an aid on which to base assumptions when planning the acceptance strategy - the ascent up and out of the V-diagram that is often neglected until too late. The acceptance strategy enables the shipbuilder to get 'off-contract' whilst ensuring that the military capability is delivered as envisaged in the ConEmp.

## 3.3 DESIGN SELECTION

It is useful to have a single comparison point when confronted with a range of design options. As the reference design embodies the requirements, many designs can be assessed against a single reference design. This is particularly valuable when the offered designs may not meet all of the requirements, such as when purchasing an existing design developed to a separate set of requirements.

Additionally, the reference design can be provided to those tendering as an example of what is desired. This can help facilitate engagement between the procuring body and those tendering, whilst also helping ensure the suitability of the tendered designs.

## 3.4 DESIGN IMPACT ASSESSMENT

### 3.4(a) Design Tailoring

Design tailoring refers to the process of modifying an existing design (such as an 'off-the-shelf' design) to meet a certain set of requirements. This is shown in Figure 2, which displays a series of Military Off-The-Shelf (MOTS) designs requiring design changes to meet the requirement set embodied by the reference design.

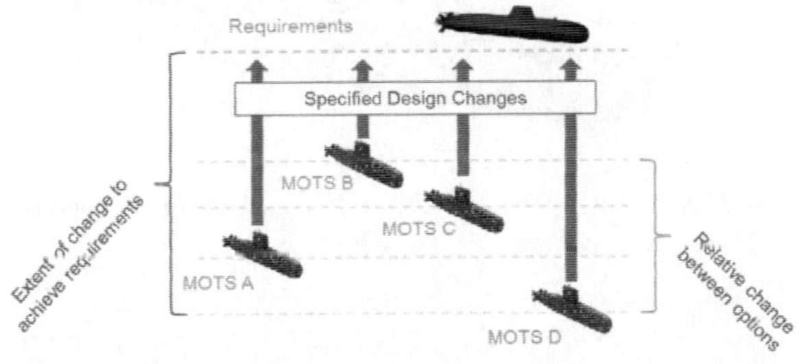


**Figure 2: Design Tailoring**

Comparison with the reference design helps identify which changes are necessary as well understanding the impact of the changes. The last point is key as the specified design changes may lead to the design becoming unbalanced and unviable and is a direct indicator of cost and risk.

As mentioned previously, the integration of the main bow sonar into the Collins Class submarine design created excessive noise due to poor hydrodynamic flow. As this was a design change specified by the procuring body (the Australian Department of Defence), then the impact of this design change could have been assessed against a reference design early in the procurement process. The issue of noise may have been identified much earlier and either avoided or mitigated before the submarine was commissioned.

### 3.4(b) Requirements Change

Due to the long timescales involved, it is not uncommon for requirements to change during a submarine's procurement. This could take a range of forms, from a relatively simple equipment change due to obsolescence, to the more complex addition or enhancement of capability. The latter was true for the Astute Class diver capability and if a reference design were available then it would have allowed the procuring body (the UK Ministry of Defence) the opportunity to independently assess the planned implementation of the design change.

Whilst implementation of design change lies with the designer, trialling capability upgrades and modifications on a reference design allows the impact to be understood prior to being specified. This allows the modification to be reviewed in light of its impact, allowing the procuring body to make truly informed decisions.

## 3.5 ACCEPTANCE STRATEGY PLANNING

As mentioned above, it is important for a procuring body to understand how compliance with the requirements they set will be demonstrated. If a reference design is available then it can be used to identify the key, measurable parameters that contribute to capability requirements. These measurable parameters can then form part of the acceptance strategy and feed into the system requirements.

## 3.6 DLOD IMPACT ASSESSMENT

The Defence Lines of Development (DLODs) encompass all the aspects necessary to deliver a capability. This is often summarised by the acronym TEPID OIL, which covers Training, Equipment, Personnel, Information, Doctrine, Organisation, Infrastructure and Logistics (Overt Defence, 2020b). Each of these areas is required to deliver a capability such as a submarine performing operations at sea.

Submarine acquisition often focusses on the equipment (the submarine) but failure to consider all of the DLODs can result in failure to deliver capability. An example is the UK's three-billion-pound procurement of the Apache attack helicopter. A shortage of trained pilots led to 11 of the 25 helicopters being mothballed for four years whilst a suitable number of pilots were trained (BBC, 2002).

As all submarine enterprises are built up, a reference design can be used as an aid early in the procurement process to identify the impact to each of the DLODs, and any associated requirements imparted by the new submarine. This allows for impact planning independent of the shipbuilder (who is unlikely to have expertise in the other DLODs), and for early identification of the drivers that will impact the other DLODs.

In some cases, the reference design may identify unacceptable impacts to other DLODs – perhaps the new submarine would require an unaffordable modification to the support infrastructure. If identified early enough then this can be fed back into the requirements of the submarine to mitigate or avoid the impact.

## 4. EXAMPLE USES

### 4.1 WYVERN SSK DESIGN

BMT produced the WYVERN® SSK concept design in 2017 to investigate the feasibility and capabilities of a small SSK (BMT, 2017). The submarine design (as shown in Figure 3) has a submerged displacement of 860 tonnes, 4 torpedo tubes and an Air Independent Propulsion (AIP) system.

**Figure 3: The BMT WYVERN® SSK Concept Design**

Whilst not a true reference design, WYVERN was an in-house design and not designed to the requirements of a specific nation. Therefore, it was necessary for BMT to develop its own set of requirements from an indicative ConEmp and hence BMT had complete control over both.

The design process allowed BMT to test and validate the requirement set, in the same way that a reference design could be used. This led to some requirements being changed when their impact became understood. One example is the specified diving depth was found to result in a high pressure hull weight for little practical gain. The requirement was ultimately reviewed, and the diving depth reduced to 210 metres.

### 4.2 BATCH 2 TRAFLAGAR CLASS

Whilst the Astute Class was impacted by the requirements change that expanded the diver capability in the design, the early part of the procurement process made use of a reference design. An initial design was produced by the UK Ministry of Defence as part of what was then known as the 'Batch Two Trafalgar Class' (B2TC) (Gates, 2018).

The initial design embodied the requirement set and included key components that were prescribed by the procuring body, such as the nuclear reactor. The B2TC reference design was an aid to design selection and directed those tendering towards an acceptable design solution, whilst also providing a 'measuring stick' to compare prospective designs against.

### 4.3 WYVERN VARIANTS

Since 2018, the WYVERN SSK concept design has become a platform for investigating new ideas. This has led to the creation of design variants to test new ideas and technologies, including an Autonomous Underwater Vehicle (AUV) hull plug (Morris & Buckton, 2018), external batteries (Ainsworth, 2018) and a variant optimised for a Coastguard role (Walchester, 2019), as shown in Figure 4.

**Figure 4: BMT WYVERN® UUV Hull Plug, External Batteries and Coastguard Variants**

Using WYVERN as a balanced baseline design which was then modified allowed for the full impact of each of the design changes to be understood. This is the same process of design impact assessment that can be done with a reference design to evaluate the impact of design tailoring or requirements change.

 

Taking the Coastguard variant as an example, the requirements were changed significantly from the baseline making some capabilities redundant whilst creating the need for new ones. The equipment and aspects of the design relating to these capabilities was changed accordingly, resulting in the AIP system and torpedo tubes being traded for a deck gun and a modified hull-form. The result of this process was a robust understanding of the design elements required to deliver the capabilities, whilst understanding the impact to the overall design.

## 5. CONCLUSIONS

A reference design is a tool that can benefit all stages of the CADMID cycle for both shipbuilders and procuring bodies. Creating a reference design early in the procurement process can help aid the generation of requirements and identify conflicts to ensure a consistent and coherent requirement set. Maintaining a reference design enables it to be used as a single point of comparison that embodies the requirements, simplifying down selection of the design options. It also can be used as an independent verification tool to assess claims made against a design. This is important for all submarine procurement programmes but particularly those that plan to procure a MOTS design.

A reference design can be used to enable the procurement body to understand the impact of their requirements. An informed decision can then be made about whether the requirement is worth the impact. This knowledge is key if specifying an aspect of the design solution such as an item of equipment.

By using a reference design, a procurement body can be fully engaged in the design process as an 'informed customer'. It allows the procuring body to look at all of the DLODs, whilst facilitating cooperation and creating a basis for dialogue between the different stakeholders.

## 6. ACKNOWLEDGEMENTS

The author would like to thank all his colleagues at BMT who have supported the writing of this paper. Special thanks go to Paul Buckton, Angus Watt, Bill Mitchelmore, Ian Duncan and Will Pope.

## 7. REFERENCES


1. UK MINISTRY OF DEFENCE (2015) *Successor submarine recruitment*, https://www.gov.uk/guidance/successor-submarine-recruitment (Accessed 27th April 2021).
2. GARCÍA-MONEDERO, J. (2019) *S-80 Plus Submarine: a cutting-edge design in continuous evolution*, Submarine Technology, Amsterdam 29th October 2019.
3. GLOBALSECURITY.ORG (2020) *Series 80 S-80 Submarine*, https://www.globalsecurity.org/military/world/europe/s-80.htm (Accessed 28th April 2021).
4. SCHANK, J.F., LACROIX F.W., MURPHY, R.E., IP, C., ARENA, M.V., and LEE, G.T. (2011a) *Learning from Experience Volume IV Lessons from the Australia's Collins Submarine Program,* RAND Corporation, ISBN: 978-0-8330-5898-0.
5. YULE, P., WOOLNER, D., (2008) *The Collins Class Submarine Story: Steel, Spies and Spin,* Cambridge University Press, ISBN: 978-0-521-86894-5.
6. SCHANK, J.F., LACROIX F.W., MURPHY, R.E., IP, C., ARENA, M.V., and LEE, G.T. (2011b) *Learning from Experience Volume III Lessons from the United Kingdom's Astute Submarine Program,* RAND Corporation, ISBN: 978-0-8330-5897-3.
7. OVERT DEFENCE (2020a) *British Defence Acquisition – A Primer on the Process*, https://www.overtdefense.com/2020/05/28/british-defence-acquisition-a-primer-on-the-process (Accessed 29th April 2021).
8. UK MINISTRY OF DEFENCE (2011) *British Maritime Doctrine*, Joint Doctrine Publication 0-10.
9. US DEPARTMENT OF DEFENCE (2001) *Systems Engineering Fundamentals*, Department of Defence Systems Management College.
10. OVERT DEFENCE (2020b) *Capability Management and the Defence Lines of Development*, https://www.overtdefense.com/2020/06/12/capability-management-and-the-defence-lines-of-development/ (Accessed 30th April 2021).
11. BBC (2002) *Army grounds Apache helicopters,* http://news.bbc.co.uk/1/hi/uk/2379501.stm (Accessed 6th May 2021).
12. BMT (2017) *WYVERN Submarine Datasheet* http://www.bmt-group.com/media/6740041/SBPD005-0517_WYVERN.pdf (Accessed 30th April 2021).
13. GATES, J. (2018) *Astute Class Nuclear Submarine Manual*, Haynes, ISBN: 978-1-7852-1071-6
14. MORRIS, D. and BUCKTON, P. (2018) *The Design Challenges of a Submarine Host AUV Hull-Plug*, Undersea Defence Technology 2018.
15. AINSWORTH, W. (2018) *Benefits and Integration Aspects of External Submarine Batteries*, Undersea Defence Technology 2018.
16. WALCHESTER, A.E., (2019) *The Coastguard Submarine: Covert Maritime Security*, Undersea Defence Technology 2019.

# NUMERICAL ESTIMATION OF DAMAGE CAUSED BY THE IMPACT OF AN EXERCISE TORPEDO ON A SUBMARINE PRESSURE HULL STRUCTURE

**R Castillo,** Chilean Navy, Chile
**N Bradbeer,** University College London, UK

## SUMMARY

Torpedo combat and evasion training exercises using trials torpedoes are valuable but involve significant risks to the target submarine in the event of an accidental impact. This paper presents a study of impacts of a nominal heavyweight torpedo against a nominal submarine pressure hull, simulated using the nonlinear explicit finite element analysis code LS-Dyna.

The model is validated against experimental work published by the Australian Defence Science and Technology Organisation, then expanded to a segment of a full-size submarine.

Impacts were simulated at torpedo speeds of 30, 40 and 50 knots. It was found that the 30 knot impact produced no plastic deformation in the pressure hull, while the 40 and 50 knot impacts produced permanent dents of several mm depth. Reference is made to DNV GL submarine structures rules to suggest that these dents will not impair operational capability.

## NOMENCLATURE

| | |
|---|---|
| DSTO | (Australian) Defence Science & Technology Organisation |
| FEA | Finite Element Analysis |
| HY-100 | A grade of high-yield steel |
| OOC | Out Of Circularity |
| SSK | Conventionally-Powered Submarine |
| UCL | University College, London |

## 1. INTRODUCTION

Realistic training is valued by the armed forces; it is widely perceived that the more realistic the training, the greater the benefit. In this context, exercise torpedo launching is considered a valuable asset, since allows to perform effective training activities for both Surface and Submarine combatants under the most realistic circumstances.

Exercise torpedoes can be launched by submarines against surface or underwater targets, and are built to be as similar to warshot torpedoes as possible; their only major difference is the absence of an explosive charge, whilst every operational (speed, range, etc.) and physical (weight, centre of gravity location, dimensions, radiated noise) aspects are typically identical.

Despite the fact the exercise torpedo does not carry an explosive load, there is still the potential to cause considerable damage if it strikes a ship or submarine at high speed. While this risk may often be considered acceptable if the target is a surface vessel, it is usually considered much higher if the target is a submarine; a loss of watertight integrity may be catastrophic on a submarine, and deformation of the pressure hull may result in a loss of strength and therefore loss of operational capability.

Besides other precautions, the main way to deal with this risk is to introduce in the torpedo a depth limit that electronically disables its capability to enter a depth zone reserved for the attacked hull. However, there is still a possibility of this this depth safety setting to fail, potentially leading to a serious accident.

Therefore, Surface Combat units train their torpedo evasion skills in a highly realistic environment. Their sonar operators are able to actually detect the generated underwater noise and the ship can train its ability to manoeuvre and evade accordingly. However, launchings against submarines are in the vast majority of cases only simulated on Fire Control System of the attacker, and broadcasted to the attacked unit, which in turn will take evasive actions. This is mainly caused by the absence of data on the potential damage that could be inflicted by the accidental impact of an exercise torpedo against a submarine hull, leading eventually to its failure with extensive human losses as a consequence.

As a negative effect, submarine training for the evading unit is performed in a very unrealistic environment when compared to a surface unit. Sonar crews are not trained in detecting the rapidly approaching incoming threat, and the submarine can perform only theoretically based evasive manoeuvres, without testing their effectiveness in any way. This last restriction is especially relevant for diesel-electric submarines, where the available energy to be spent evading is crucial to the unit's potential survivability against enemy attack.

This report details work conducted as part of the first author's MSc studies at UCL to understand the potential damage to a submarine pressure hull resulting from an accidental strike from an exercise torpedo, with the intention of better informing the decision making process around risk management of torpedo firing exercises against submarines. A series of finite element models were developed, first replicating a series of physical experiments conducted by the Australian Defence Science & Technology Organisation, and then expanding into a full-scale submarine hull section. The explicit nonlinear FE solver LS-Dyna was used to assess the behaviour of the hull during and after impact.

## 2. LITERATURE REVIEW

The modern submarine pressure hull design process is still frequently based on work done by Kendrick [1], which treat pressure hull collapse as a complex interaction between yielding and buckling, in both interframe and overall modes. Overall failure in particular is very sensitive to out-of-circularity (OOC) flaws in the cross section of a cylindrical pressure hull; Lu & Yu [2] demonstrated using energy methods that OOC arising from accidental dents can lead to early precipitation of collapse.

Due to the practical difficulties of conducting physical impact experiments, the studies described in this paper were conducted using numerical methods. The tool used was LS-Dyna, an explicit nonlinear finite element analysis code particularly suitable for use in modelling collisions. LS-Dyna is widely used in the automotive industry for modelling the effect of accidents [3], as well as in military contexts for modelling the impact of explosions and high speed projectiles.

LS-Dyna has been used to simulate naval collisions, which are complicated by the introduction of hydrostatic and hydrodynamic forces. Donner et al [4] described a coupled model using LS-Dyna and the dynamics simulation code MCOL to simulate collisions between a ship and a submarine surfaced or at periscope depth, which was verified against large scale experiments conducted for the French navy. The use of a coupled model allowed the large scale dynamics of the system to be explored, including body motions after the impact, and the results from LS-Dyna were used to predict the residual strength of the submarine hull post-collision.

In the case of a submarine being struck by an exercise torpedo, it may not be necessary to use a fully coupled dynamics model; the motions of the bodies after impact are of little interest compared to the deformation and stresses arising in the pressure hull from the impact. Considerable work has been conducted by the Royal Australian Navy in this area, with Chatterton & Ojeda [5] conducting numerical modelling of blank torpedo impacts against a section of pressure hull, which were validated against physical drop tests as reported by Van der Schaaf et al [6]. This work concluded that numerical analysis can deliver a usefully accurate simulation of torpedo/pressure hull impacts. However, Chatterton & Ojeda's work replicated drop tests conducted in air against small panel sections, and furthermore did not consider the effects of hydrostatic pressure acting on the hull during and after the impact event.

MacKay [7] provided a useful review of the state of the art in pressure hull design and analysis, which included several key recommendations for numerical studies of pressure hulls, in particular the use of nonlinear solvers, 4-8 node shell elements and careful choice of material models.

## 3. METHOD

The work conducted in this study was designed in three main stages, aiming to ensure the accuracy of numerically calculated responses and develop a validated full-scale model to be trialled.

The first stage's validation studies replicated Trials 6 & 2 from Van der Schaaf's [6] experiments. Trial 6, a drop onto an unstiffened steel plate, was the only one of the drop tests that was reportedly modelled with FEA software. Replication of this experiment allowed validation of the LS-Dyna model against an existing validated numerical solution with similar results data available [5]. Once this initial model had demonstrated that satisfactory model parameters were in use, a more complex stiffened arrangement was developed based on Van der Schaaf's Trial 2. This allowed for validation of the stiffened model.

In the second stage, torpedo impact simulations, a larger stiffened model was developed, corresponding to a single frame bay of a typical SSK structure at full scale. This model was run with two pressure hull materials (mild steel and HY-100 steel) and three torpedo impact velocities (30, 40 and 50 knots.) This stage provided reformation and stress values which were then used to evaluate the criticality of the inflicted damage.

The final stage was assessment of the hazards to the submarine. Transient stresses during the impact were measured and used to assess the risk of the submarine undergoing collapse immediately due to the impact. The permanent set in the structure following the impact was then used to assess the impact on the hull's strength and whether any subsequent operational restrictions on depth would be required. Conclusions were then drawn from these to advise the operator on safe trials conditions.

## 4. VALIDATION STUDIES

Models were developed in LS-Dyna to replicate Van der Schaaf's Trial 6 and Trial 2 [6]. Both of these trials had used the same laboratory setup, as shown in Figure 1. This rig dropped a carriage with an aluminium model of a torpedo nose a distance of 2m vertically onto a curved plate that was bolted along the long edges to simulate the restraints on a submarine pressure hull section.

Trial 6 used an unstiffened 3mm steel plate. In Trial 2 the 3mm plate was stiffened with two ring frames.

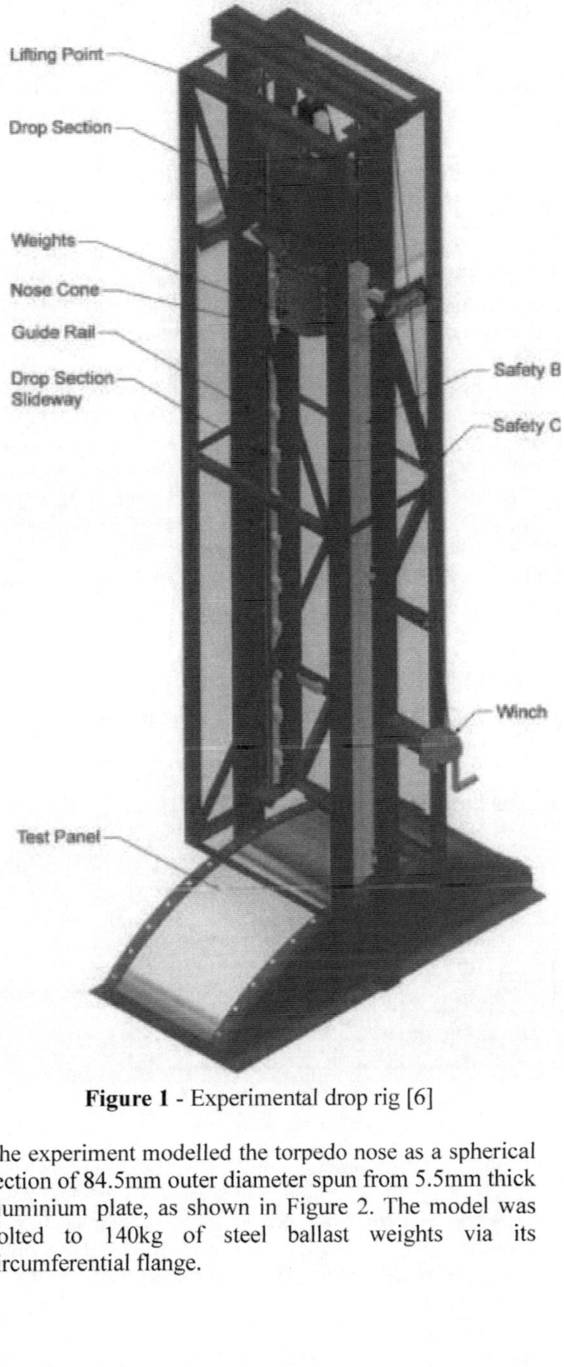


**Figure 1** - Experimental drop rig [6]

The experiment modelled the torpedo nose as a spherical section of 84.5mm outer diameter spun from 5.5mm thick aluminium plate, as shown in Figure 2. The model was bolted to 140kg of steel ballast weights via its circumferential flange.

**Figure 2** - Torpedo nose model [6]

The torpedo nose was modelled in LS-Dyna as a spherical shell with 3,750 quad shell elements and a weight of 140kg distributed over the nodes of the upper half.

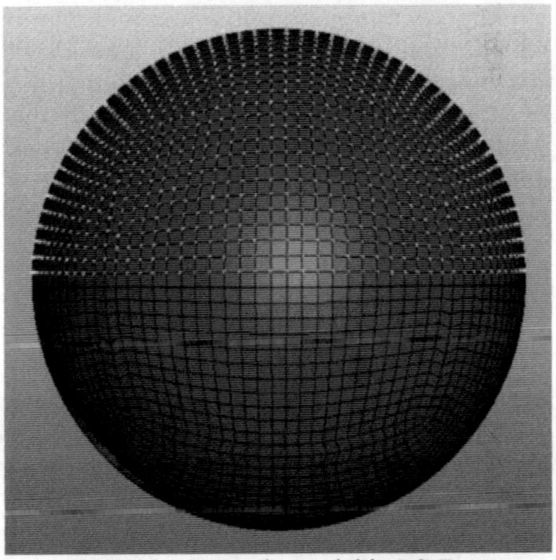

**Figure 3** – Torpedo model in LS-Dyna

For Trial 6, the unstiffened 3mm plate was modelled using 4-node Belytschko-Tsay quad shell elements, with sides of 5mm. 36,100 elements were required to model the entire plate.

For Trial 2, the stiffened 3mm plate was modelled as for Trial 6, with the addition of two ring stiffeners. These were modelled using the same 4-node Belytschko-Tsay quad shell elements, for a total model size of 50,547 elements.

Two materials were used, both using the MAT_024 Piecewise Linear Plasticity model [8], mild steel for the plating and stiffeners, and aluminium for the torpedo nose sphere. Material parameters were as shown in Tables 1 and 2. The MAT-024 model for steel uses a piecewise interpolation model for the stress/strain relationship post yield defined by the parameters EPS1-3. Each of these specifies a combination of plastic strain and stress which define the shape of the post-yield stress-strain curve.

**Table 1**. Material properties for validation model steel

| Parameter | Value |
|---|---|
| Density | $7.850\ te.m^{-3}$ |
| Elastic modulus | $210\ GPa$ |
| Poisson's Ratio | 0.3 |
| Yield Stress | $380\ MPa$ |
| EPS1 | 0 / 380 |
| EPS2 | 0.004 / 520 |
| EPS3 | 19.9 / 530 |

**Table 2**. Material properties for validation model aluminium

| Parameter | Value |
|---|---|
| Density | $2.79\ te.m^{-3}$ |
| Elastic Modulus | $69\ GPa$ |
| Poisson's Ratio | 0.32 |
| Yield Stress | $80\ MPa$ |
| ETAN | 1600 |

Boundary conditions were set with all nodes on the plate's long edges constrained in all six degrees of freedom, while the nodes on the short edges were left unconstrained. This emulated the experimental rig's boundary mounting arrangement where the long edges were bolted and the short edges left free. The effect of these constraints can be seen in the stress distribution shortly after impact on the unstiffened plate, in Figure 4.

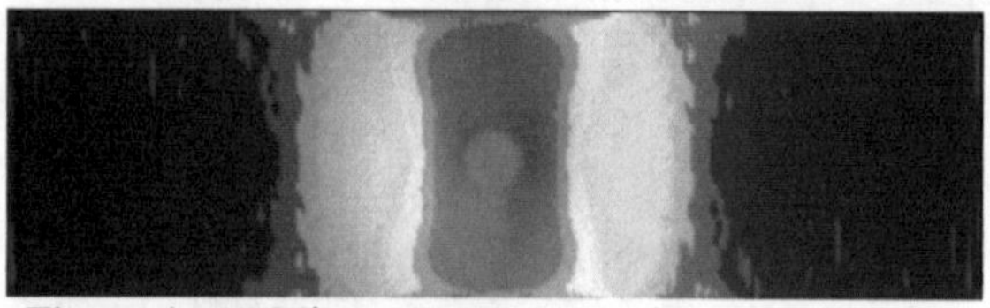

**Figure 4**. von Mises stress contour plot across the plate shortly after impact

A timestep of $8 \times 10^{-7}$ seconds was used, terminating the simulation after 0.03 seconds.

The radial permanent set was measured along the axial and circumferential lines passing through the centre of the impact point. The deflection was nondimensionalised as w⁄t where w was the radial deflection and t the plate thickness. The locations were expressed as nondimensionalised distance from the centre of the impact point as x⁄r and y⁄r respectively, where r was the torpedo radius, x the axial distance (parallel to the short edge) and y the circumferential distance (parallel to the long edge). The deflections were compared to the DSTO experimental data. Table 3 shows the absolute errors comparing DSTO's ANSYS model and this study's LS-Dyna model to the experimental data.

**Table 3.** Comparison of deflections in LS-Dyna model with DSTO Experiment and ANSYS model [5], [6]

| Position y/r | w/t DSTO Experiment | w/t DSTO ANSYS | ANSYS Absolute Error (%) | w/t LS-Dyna | LS-Dyna Absolute Error (%) |
|---|---|---|---|---|---|
| 1.0 | -9.5 | -7.1 | 25.26 | -7.5 | 20.70 |
| 0.75 | -10.2 | -8.9 | 12.75 | -9.4 | 7.84 |
| 0.5 | -11.1 | -10.5 | 5.41 | -11.2 | 0.90 |
| 0.25 | -11.5 | -10.9 | 5.22 | -11.7 | 1.74 |
| 0.0 | -11.5 | -11.2 | 2.61 | -11.7 | 1.74 |
| -0.25 | -11.4 | -10.9 | 4.39 | -11.7 | 2.63 |
| -0.5 | -11.2 | -10.5 | 6.25 | -11.2 | 0.00 |
| -0.75 | -10.4 | -8.9 | 14.42 | -9.4 | 9.62 |
| -1.0 | -9.4 | -7.1 | 24.47 | -7.5 | 19.86 |

It may be observed in Figure 5 that while both the ANSYS and LS-Dyna models underpredict the deflection at the outer edges of the impact region by 20-25%, they model the central deflection very well, with errors of only 1-3%. The LS-Dyna model's errors in the longitudinal direction were universally smaller than the ANSYS model's errors. It may be seen from Figure 5 that both numerical models produced similar deflected shapes, which were both more tightly curved than the experimental case.

In the transverse direction (Figure 6) the two numerical models again produced similar shaped distortions, with the ANSYS model again predicting lower displacements than the LS-Dyna model. However, in this case both numerical models overpredicted the displacement at the edged of the impact region.

These results were considered to be good matches for the DSTO data, indicating that the LS-Dyna model was set up with appropriate parameters.

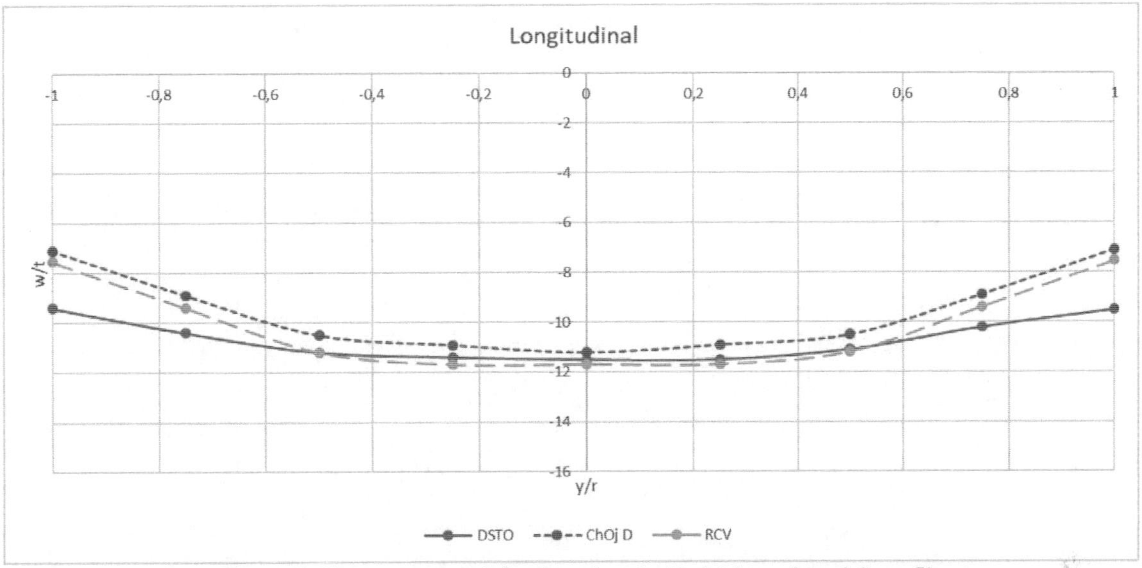


**Figure 5**. Plate permanent deflections, longitudinal (circumferential) profile

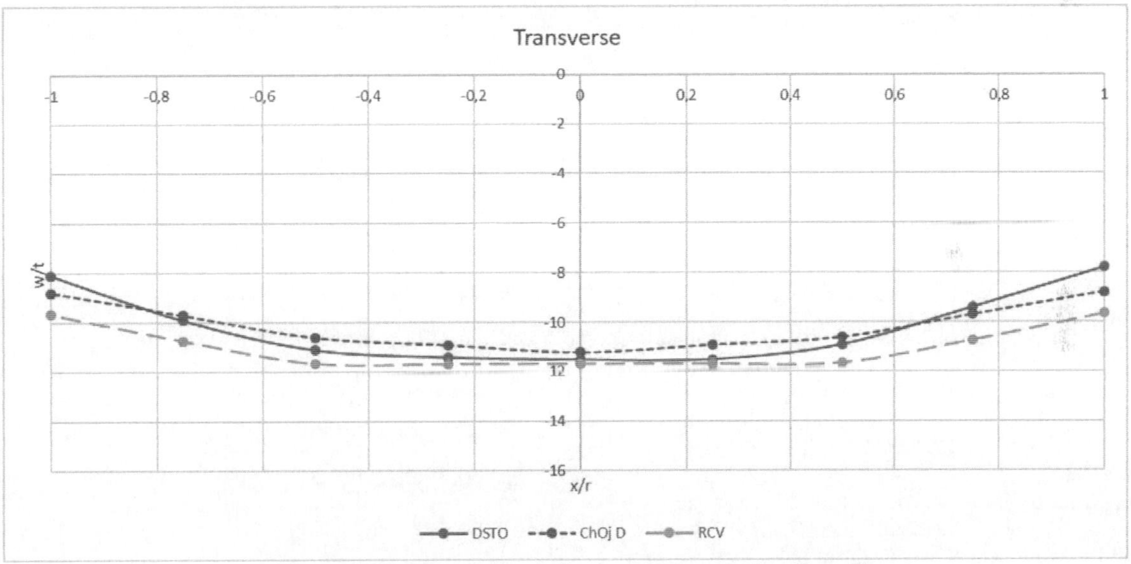


**Figure 6**. Plate permanent deflections, transverse (axial) profile.

This simulation was repeated with the Trial 2 stiffened plate geometry.

The effect of stiffening on the dent size and shape was profound, as shown in Figure 7, comparative plots of w/t against x/y and y/r for the stiffened and unstiffened cases. Not only was the dent considerably shallower in the stiffened plate, the horizontal size of the dent was constrained significantly by the ring stiffeners. It can be seen in the transverse (axial) plot that the ring frames effectively constrained the plating with very little deflection and some rotation of the plating around their roots. The dent is similarly narrower in the longitudinal direction, probably due to shear effects in the plate.

No detailed experimental data was available for comparison; the only available data point was the depth of the dent in the plating, which the LS-Dyna model matched within 10%. This was considered acceptable for validation.

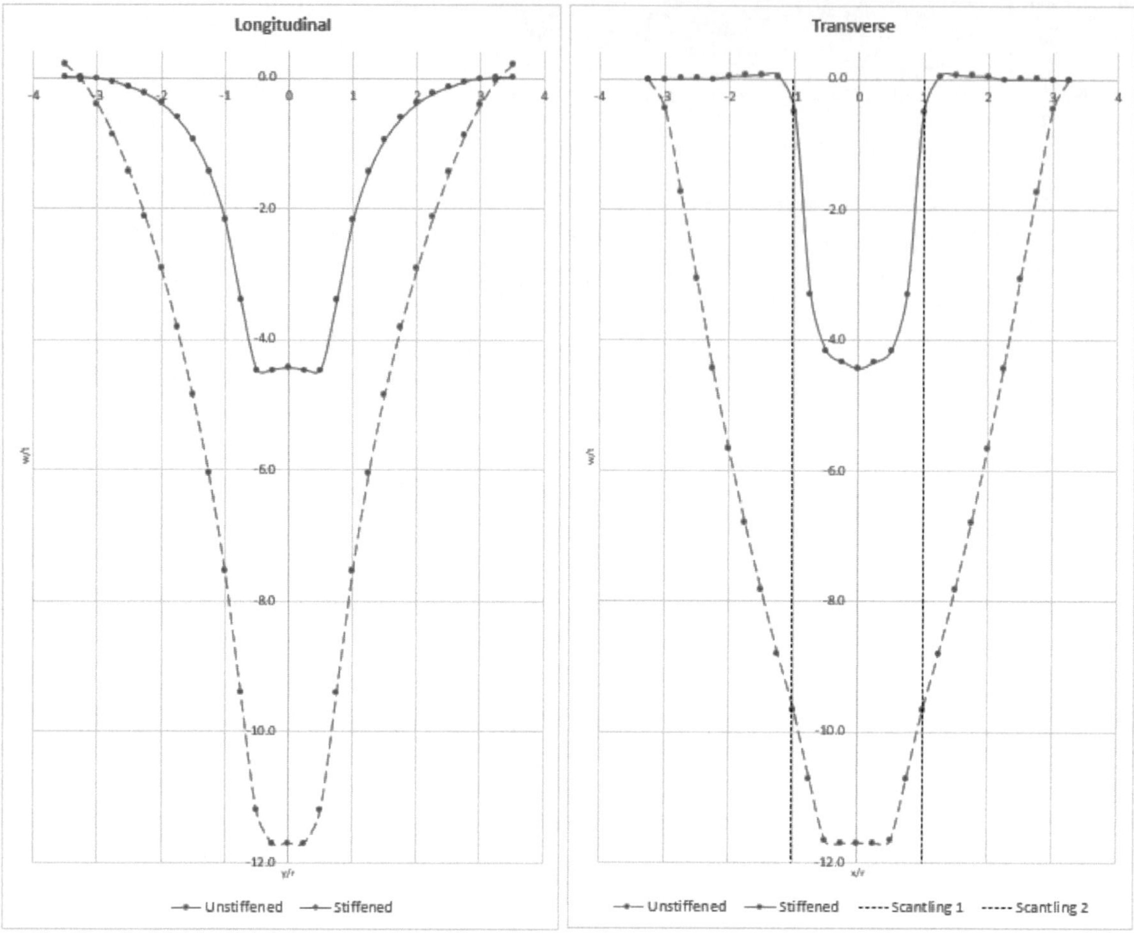


**Figure 7.** Comparative permanent deflection profiles in unstiffened and stiffened plates

## 5. TORPEDO IMPACT SIMULATIONS

A full-scale LS-Dyna model of a partial pressure hull was developed, using nominal unclassified structural parameters representative of a typical SSK. Shots were run using mild steel and HY-100 steel to gauge the effect of material strength on the permanent deformation, and then a series of shots were run against an HY-100 hull with torpedo impact speeds of 30, 40 and 50 knots.

**Table 4**. Pressure hull geometric parameters

| **Geometric Parameter** | **Value** |
|---|---|
| Pressure hull external radius | $3.07m$ |
| Shell plating thickness | $30mm$ |
| Unsupported interframe distance | $0.58m$ |
| Frame cross sectional area | $7.185 \times 10^{-3} m^2$ |
| Frame web thickness | $12mm$ |

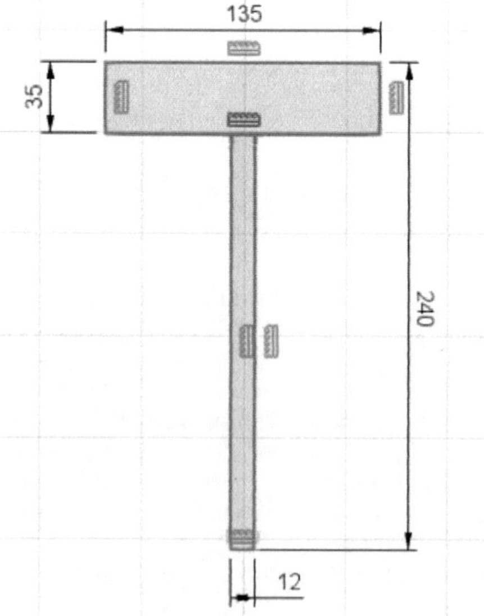


**Figure 8**. Ring frame cross section

**Table 5.** Mild Steel material properties

| Parameter | Value |
|---|---|
| Density | $7.850\ te.m^{-3}$ |
| Elastic Modulus | $207\ GPa$ |
| Poisson's Ratio | 0.3 |
| Yield Stress | $305\ MPa$ |
| EPS1 | 0 / 305 |
| EPS2 | 0.004 / 420 |
| EPS3 | 19.9 / 430 |

**Table 6.** HY-100 material properties

| Parameter | Value |
|---|---|
| Density | $7.850\ te.m^{-3}$ |
| Elastic modulus | $207\ GPa$ |
| Poisson's Ratio | 0.3 |
| Yield Stress | $700\ MPa$ |
| EPS1 | 0 / 700 |
| EPS2 | 0.004 / 880 |
| EPS3 | 19.9 / 890 |

The model was not subjected to hydrostatic pressure during the impact; this was considered by later calculation. Since the submarine plating was thick relative to the expected size of the impact deformation features, thick shell elements were used to gain improved accuracy. 2D shell elements were retained for the ring frame geometry since the validation tests showed they suffered almost no deformation; this was in line with best practices for FEA modelling of submarine geometry recommended by MacKay [7].

The scope of the model was limited to a region of the pressure hull of similar size to those used for the DSTO drop trials. The validation runs had shown that neither the stress fields nor the deformations in the stiffened plate had approached any of the edges, so this model seemed likely to be large enough to avoid edge effects.

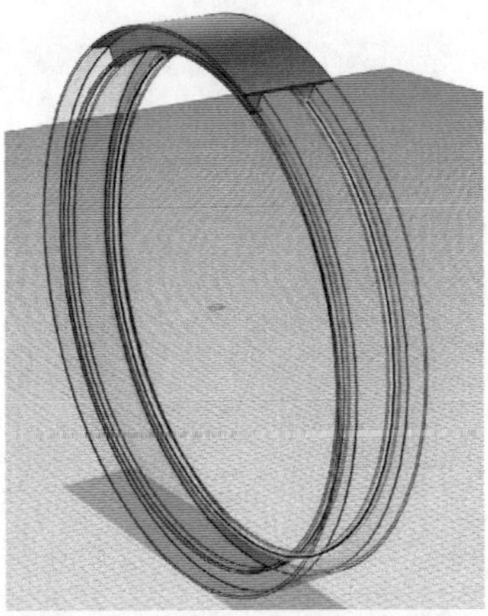

**Figure 9.** Region of pressure hull modelled in LS-Dyna (with remainder of hull shown in wireframe)

Plate boundary conditions were set to be the same used for the validation trials, with the long edges clamped in all degrees of freedom, and the sort edges unconstrained.

The impacting torpedo was modelled as a sphere of aluminium 533mm in diameter, corresponding to the size of a typical heavyweight torpedo. While this was quite a broad approximation of the geometry and material it was deemed acceptable since the torpedo head was expected to deform more readily than the pressure hull shell plating and, in any case, the deformations of the torpedo head were not the subject of interest.

The inertia of the torpedo was simulated by distributing 1.5te of mass over the nodes comprising the rear half of the sphere. Sensitivity analysis showed the results to be quite sensitive to the choice of mass distribution, and further investigation is recommended.

Each of these models were run for 0.11s with a timestep of $8 \times 10^{-8}s$.

The variation in material strength from Mild Steel to HY-100 had a very significant effect on the plastic deformation in the hull plating, although it must be remembered that these were two structures with matched geometry not matched collapse strength. Orthogonal profiles through the impact dent of a 50-knot torpedo are shown in Figures 10 and 11, where it can be seen that the dent in the mild steel structure is around five times deeper than the one in the HY-100. While this is not unexpected, it highlights the importance of material model choice, and illustrates that it would be useful to have experimental data from testing plates in HY-80 or HY-100.

Three impacts against the HY-100 structure were simulated, with torpedo impact speeds of 30, 40 and 50 knots. The resulting permanent set is shown in Figure 12 along the transverse profile through the impact centre, with the ring frame centrelines shown.

The 30 knot impact left no perceptible plastic deformation, while the 40 knot impact left a dent 7% of the plating thickness deep and the 50 knot impact left a dent 25% of the plating thickness deep.

**Table 7.** Permanent set depth from torpedo impacts

| Torpedo impact speed | Permanent set depth (mm) | Permanent set depth $(^{w}/_{t})$ |
|---|---|---|
| 30 knots | 0.00mm | 0.00 |
| 40 knots | 1.91mm | 0.06 |
| 50 knots | 7.77mm | 0.26 |

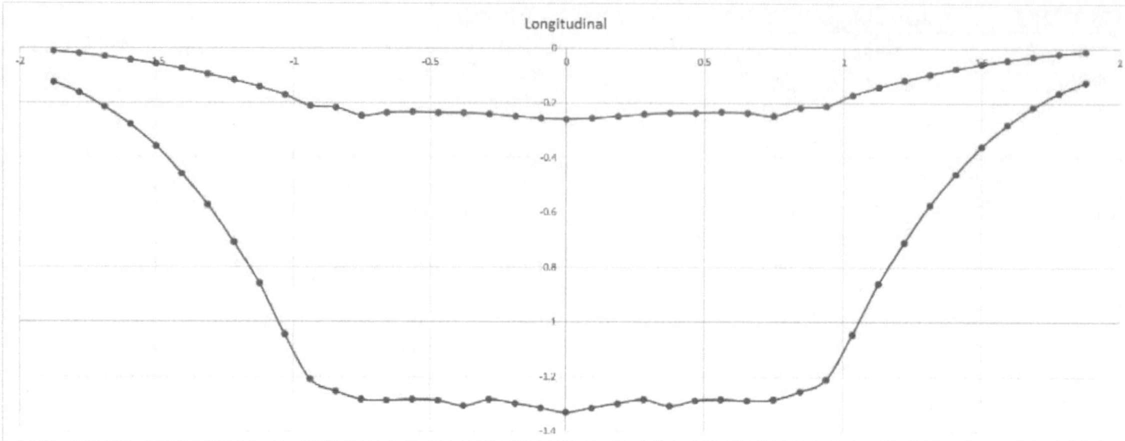


**Figure 10**. Comparative permanent set in mild steel and HY-100 models, longitudinal (circumferential) profile

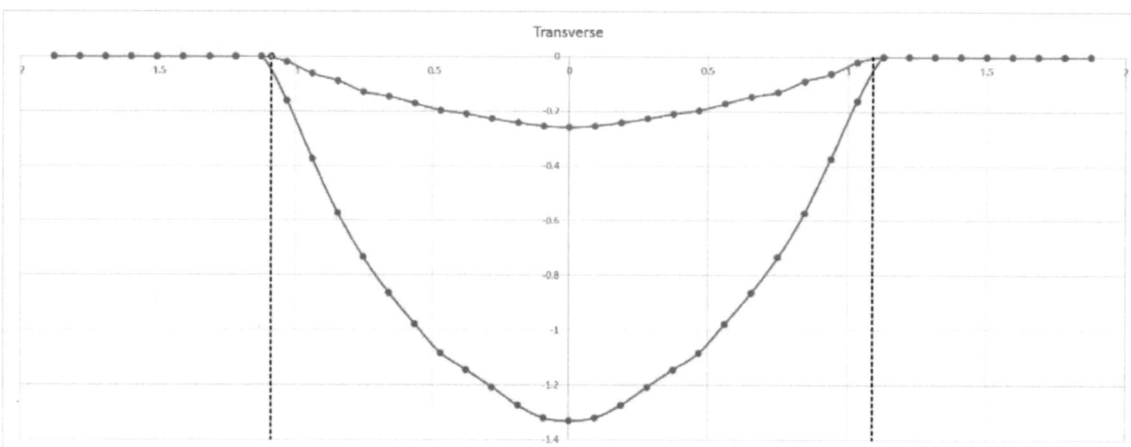


**Figure 11**. Comparative permanent set in mild steel and HY-100 models, transverse (axial) profile

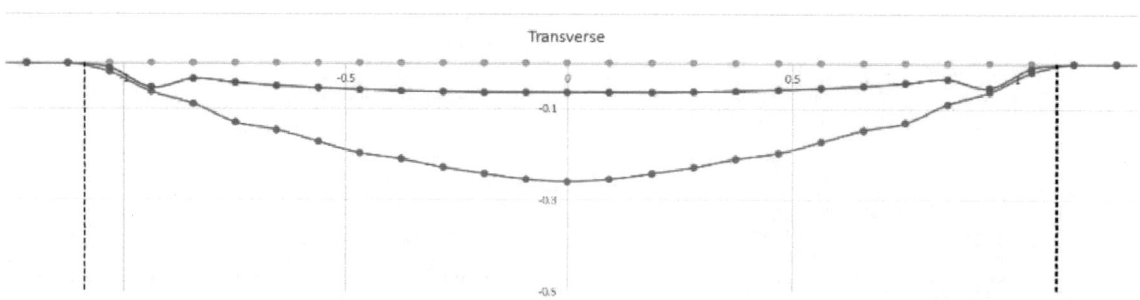


**Figure 12**. Comparative permanent set in HY-100 steel from 30, 40 and 50 knots impacts

## 6. HAZARD ASSESSMENT

The effect of the torpedo impact on the submarine was considered in two ways; the immediate threat of pressure hull collapse and the longer term reduction in strength that might require a reduction in operational diving depth to assure crew safety.

The strength of the pressure hull against collapse was calculated using the methods derived from Kendrick's work [1], [9]. This recognises the critical regions where the onset of yielding will first occur as the plating at the root of the ring frame ($\sigma_7$) and the plating midway between two frames ($\sigma_3$ in the outer fibre of the plating and $\sigma_5$ at the mid-thickness of the plating.) Yielding will first be reached at $\sigma_7$, followed by $\sigma_3$. However, the plastic region will grow more quickly at the midpoint of the frame bay, so the critical stress is usually considered to be $\sigma_5$. Once this reaches yield, it indicates that the mid frame bay plastic region has extended significantly through the plating.

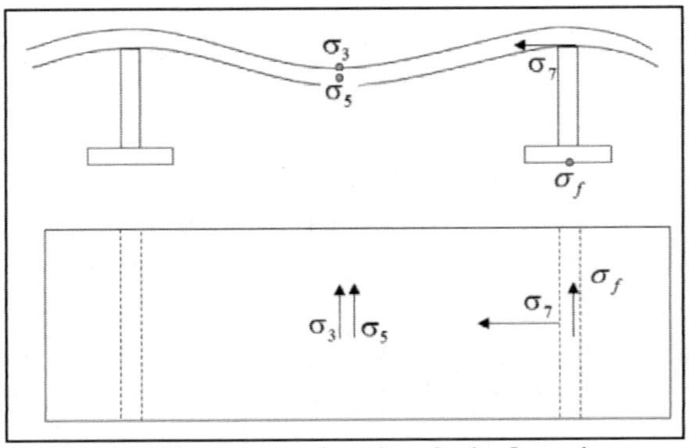


**Figure 13**. Key stress locations in the frame bay

Formulas from Kendrick may be used to calculate these stresses arising from hydrostatic pressure at any given depth of water.

The additional stresses arising from the impact of the torpedo were obtained by monitoring von Mises stress in the plate elements throughout the impact process. As may be seen from the stress contour plot in Figure 14, the peak stresses occurred coincident with $\sigma_7$ and $\sigma_5$.

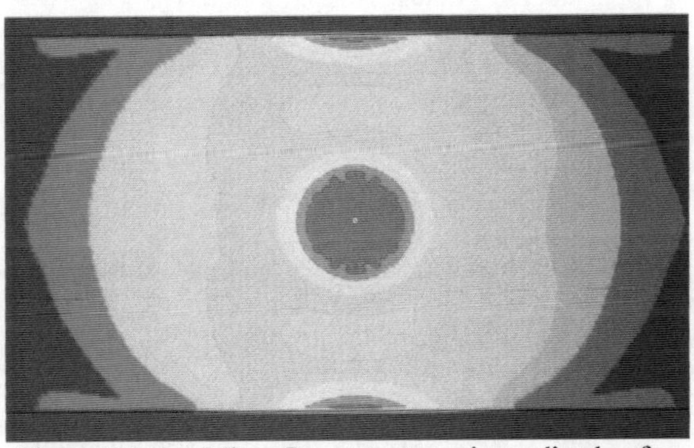

**Figure 14**. von Mises Stress contours immediately after torpedo impact

The criteria for interframe collapse presumes that collapse occurs when $\sigma_5$ reaches some fraction of yield stress, as influenced by the effects of buckling. Since plastic deformation of the whole plate thickness occurred in the 40 knot and 50 knot impacts, that criteria had clearly been failed in both those runs. To determine the risk of collapse occurring during a fast impact would require finite element analysis with external pressure loads applied. This was outside the scope of this study, since that calculation would be better suited to an implicit FEA solver than the explicit solver used to resolve this high-speed impact.

No plastic deformation occurred in the 30 knot impact, however, so some margin of strength existed before the plastic region around the mid frame bay reached the full plate thickness. This margin could be used to support some hydrostatic load, giving a depth at which the impact would cause plastic deformation.

**Table 8**. Von Mises stresses at critical locations induced by torpedo impact at different speeds

| | 50 knots | 40 knots | 30 knots |
|---|---|---|---|
| Dynamic $\sigma_5$ stress | 819 MPa | 585 MPa | 223 MPa |
| Dynamic $\sigma_7$ stress | 758 MPa | 716 MPa | 601 MPa |

Noting that the HY-100 steel had a yield stress of 700MPa, there is a margin of 99MPa before $\sigma_7$ will reach yield. Using Kendrick's equations, a hoop stress of 99MPa will be generated at $\sigma_7$ when the boat is at a depth of 109m. Therefore the 30 knot impact might be expected to put no part of the shell plating into yield if it occurs at a depth of less than 100m. However, it must be stressed that collapse behaviour is complex, and to have confidence in this value, further FE modelling incorporating external pressure loads should be conducted.

To consider the long-term effects of dents arising from higher speed impacts, the dents were treated as Out of Circularity features, according to DNV GL's structural rules for submarines [10]. While this method is normally applied to frames, using sample points spaced approximately every metre around the submarine, the method is stated to be applicable to the shell plating and can be applies with a greater density of sample locations. In this case 96 sampling locations were used, giving a sample spacing of 0.1m, fitting 10 sampling locations into the width of the dent caused by the 50 knot impact.

The DNV GL rules allow an assessment to be made of OOC features on mid-term and long-term performance. In this case, the deformations resulting from both the 40 knot and 50 knot impacts passed the DNV criteria, allowing unrestricted operation following both impacts.

## 7. CONFIDENCE

Choice of model parameters is of great importance when conducting finite element modelling. In this case, the model was found to replicate experimental data well, which gives confidence. The boundary conditions were selected to match the DSTO physical trials, and there were no significant deformations observed around the boundaries of the model, suggesting that the model was not undersized. Low stresses and deformations in the plating outside of the two ring frames suggest that there was no need to incorporate more rings frames into a longer model.

The choice of material model is also important. Again, results which well matched the DSTO trials data are encouraging that MAT_024 was an appropriate choice, with appropriately selected parameters.

In summary, the authors have good confidence in the FEA model and the resulting stress and plastic deformation results.

Confidence is lower in the models for the implications on whole-boat strength. Some preliminary estimates have been made, but this would benefit from additional finite element modelling, most likely by exporting the deformed node positions into an implicit solver to explore the damaged structure's response to external hydrostatic load.

## 8. CONCLUSIONS

A finite element model was constructed which replicated the DSTO physical drop tests well.

The expanded model was then used to simulate impacts from a heavyweight torpedo at 30, 40 and 50 knots against the plating directly between two frames, of a nominal submarine whose pressure hull comprised 30mm of HY-100 supported by ring frames every 550mm.

When modelled without hydrostatic pressure, the 30 knot impact left no permanent set in the plate, indicating that the material had remained within the elastic regime. It is assessed that the plate would remain elastic for 30 knot impacts down to a depth of 100m.

The 40 and 50 knot impacts left localised permanent deformation in the shell plating, which was largely contained by the adjacent ring frames. The resulting dents were approximately 2mm deep for the 40 knot impact and 8mm deep for the 50 knot impact. The latter corresponds to 0.27% of the pressure hull radius, which is below the typical threshold for acceptable Out of Circularity.

The DNV-GL submarine structural rules were used to assess the deformations arising from the 40 and 50 knot impacts and found them to give no impact on long term operational performance.

Further FE analysis of this situation is warranted, particularly in adding hydrostatic loading. This might be done by linking the results of the explicit LS-Dyna model to an implicit FE solver, which may be less computationally intensive than adding a hydrostatic load into the explicit model.

## 9. REFERENCES


1. KENDRICK, S. B., Collapse of Stiffened Cylinders Under External Pressure. Proc. Conf. on Vessels under Buckling Conditions, Inst. Mech. Eng, 1972

2. LU, G. & YU, T., Energy Absorption of Structures and Materials. Elsevier, 2003

3. CHUNKE, L., XINPING, S. & JIAO, W. Simulation Analysis of Car Front Collision Based on LS-Dyna and Hyper Works. J. Transportation Technologies, 2014

4. DONNER, R., BESNIER, F. & LE SOURNE, H. Numerical Simulation of Ship Submarine Collisions. Proc. PRADS, 2001

5. CHATTERTON, M. & OJEDA, R. Finite Element Simulation of Blank Torpedo Impact on Submarine Hulls. Ocean Engineering, issue 19, 2014

6. VAN DER SCHAFF, P. Small Scale Drop Tower Test for Practice Torpedo Impact Modelling. DSTO, 2012

7. MACKAY, J. Structural analysis and design of pressure hulls: the state of the art and future trends. Technical Memorandum, Defence R&D Canada-Atlantic, 2007

8. Livermore Corporation Software Technology. LS-Dyna Keyword Users Manual, Vol II. Livermore, 2014

9. Bradbeer, N. MSc Naval Architecture Subsea Structures Course Notes. UCL, 2019

10. DNV GL. Rules for Classification of Naval Vessels, Part 4, Sub Surface Ships. DNV, 2015


## 10. AUTHORS' BIOGRAPHY

**Cdr Rafael Castillo** is an officer of the Chilean Navy. The main body of this work was undertaken as part of his studies for a MSc in Naval Architecture at UCL.

**Dr Nick Bradbeer** is a member of the Royal Corps of Naval Constructors, currently posted as an honorary Associate Professor of Naval Architecture at UCL, where he teaches ship and submarine structures.

The views expressed in this paper are the authors' own and do not necessarily represent the views or position of their employers.

# THE VERTICAL FORCE REQUIRED FROM THE AFT CONTROL SURFACES TO ENABLE A SUBMARINE TO MAINTAIN A LEVEL TURN

**H Kim**, Australian Maritime College, Australia
**D Ranmuthugala**, Defence Science and Technology Group, Australia
**M R Renilson,** Pacific ESI, Australia

## SUMMARY

When a submarine turns in the horizontal plane it experiences a complex flow around the hull, casing and sail. This results in out-of-plane loads in heave and pitch. To design the Aft Control Surfaces (ACS) such that adequate force can be applied to counteract these out-of-plane loads, it is necessary to be able to accurately predict the magnitude of these forces during the turn.

In this paper, predictions of the BB2 generic submarine with X-plane ACSs turning in the horizontal plane using a Coefficient Based Model (CBM) were compared with those from a fully coupled Computational Fluid Dynamics (CFD) model, and results from free running model tests.

The CFD predicted the measured ACS angles well, whereas the CBM predicted that much larger angles were required. This is possibly due to the CBM not accurately predicting the out-of-plane loads on the turning submarine. Further work is required to investigate this issue.

## NOMENCLATURE

*Abbreviations*

| | |
|---|---|
| ACS | Aft control surface |
| CB | Centre of buoyancy |
| CG | Centre of gravity |
| CBM | Coefficient based model |
| CFD | Computational fluid dynamics |
| DOF | Degrees of freedom |
| FRM | Free running model |
| LCB | Position of longitudinal centre of buoyancy (from the nose tip) |
| LCG | Position of longitudinal centre of gravity (from the nose tip) |
| RBD | Rigid Body Dynamics |
| SSK | Conventional attack submarine |
| URANS | Unsteady Reynolds Averaged Navier Stokes |
| VCB | Position of vertical centre of buoyancy (from the keel) |
| VCG | Position of vertical centre of gravity (from the keel) |

*Symbols*

| | |
|---|---|
| $B$ | Breadth |
| $D$ | Depth from deck |
| $D_{sail}$ | Depth from sail tip |
| $Ixx$ | Moment of inertia about $x$ axis |
| $Iyy$ | Moment of inertia about $y$ axis |
| $Izz$ | Moment of inertia about $z$ axis |
| $L$ | Length of submarine |
| $m$ | Mass of submarine |
| $r$ | Yaw velocity |
| $r'$ | Non-dimensional yaw velocity (= $rL/U$) |
| $U$ | Velocity of origin of body axes (= $\sqrt{u^2+v^2+w^2}$) |
| $v$ | Sway velocity |
| $v'$ | Non-dimensional sway velocity (= $v/U$) |
| $x'$ | Non-dimensional surge displacement (= $x/L$) |
| $y'$ | Non-dimensional sway displacement (= $y/L$) |
| $\theta$ | Pitch angle |
| $\rho$ | Water density |
| $\delta_r$ | Effective rudder angle |
| $\delta_s$ | Effective stern plane angle |
| $\delta_{SBup}$ | Upper starboard side ACS angle |
| $\delta_{SBdn}$ | Lower starboard side ACS angle |
| $\delta_{Pup}$ | Upper port side ACS angle |
| $\delta_{Pdn}$ | Lower port side ACS angle |
| $\delta_{sailplanes}$ | Sailplane angle |

## 1. INTRODUCTION

When a submerged submarine turns in the horizontal plane it experiences a complex flow around the hull, casing and sail. This results in out-of-plane loads in heave and pitch as discussed in [1]. To design the Aft Control Surfaces (ACS) such that adequate force can be applied to counteract these out-of-plane loads, it is necessary to understand their magnitude.

The motion of a submarine can be predicted through simulations employing a Coefficient Based Model (CBM) [2] or a fully coupled rigid body model using Computational Fluid Dynamics (CFD) to predict the complex flow around the submarine [3]. However, the latter model requires significant computational resources and is not ideal for the early concept phase of a submarine design.

In order to understand the limitations of a CBM for predicting the turning of a submarine with an X-plane Aft Control Surface (ACS) configuration, the results from CBM simulations are compared in this paper with those using a fully coupled CFD simulation, and those from a Free Running Model (FRM) experiment.

## 2. BACKGROUND

This study utilised the BB2 generic submarine geometry [4], which represents a typical conventional (SSK) submarine. The BB2 submarine geometry consists of a sail with sailplanes, ACS is an X-plane configuration, and a hull with a casing. This study employed the model scale BB2 geometry (Figure 1) based on Froude scaling laws with a scale factor of 18.348, giving the model a length (*L*) of 3.826m.

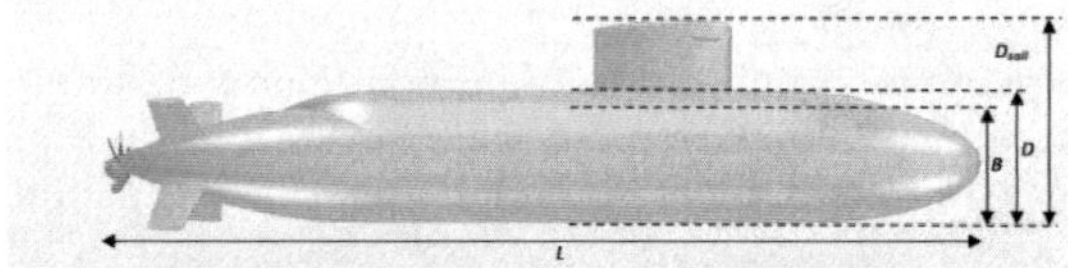

Figure 1. The BB2 submarine geometry

The submarine motion is described by a 6-DOF body coordinate frame of reference (see Figure 2), in which the origin (O) is located at the Centre of Gravity (*CG*) with the positive directions along the *x*, *y* and *z* axes being forward, starboard, and vertically downwards respectively.

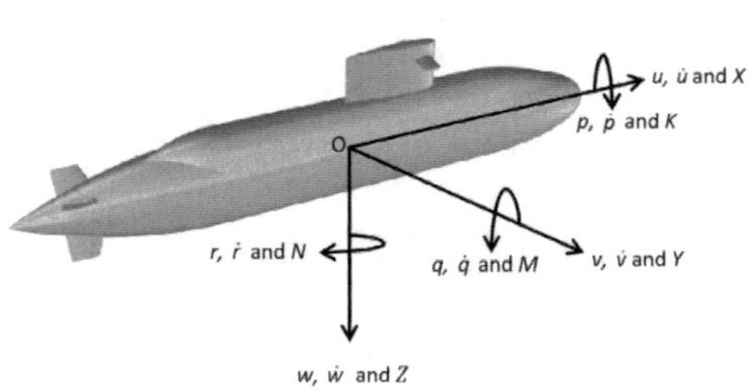


Figure 2. Body fixed coordinate with the origin (O) located at the Centre of Gravity (*CG*) of the BB2

The Centre of Buoyancy is at the same longitudinal position as the Centre of Gravity and the vertical stability lever was set at 0.022 m with the *CG* below the Centre of Buoyancy (*CB*). This gives a vertical stability lever of 0.0057 × *L*. The details of the mass properties are presented in Table 1.

The X-plane configuration consisted of four control planes moving independently to maintain translational and rotational stability in six degrees of freedom (6-DOF) throughout the course of the manoeuvres. The autopilot commanded effective rudder ($\delta_r$) and effective sternplane ($\delta_s$) angles, with individual plane angles ($\delta_{Pup}$, $\delta_{Pdn}$, $\delta_{SBup}$, and $\delta_{SBdn}$) calculated using Equations (1) to (4) based on the right hand rule with the thumb pointing away from the body [4]. The sailplane angle is calculated using Equation (5), with the angle being positive for trailing edge down when looking from starboard side. Note that the maximum plane angle for each of the ACSs was limited to 30 degrees.

Table 1. Mass properties of the model scale BB2 submarine

| Dimension | Value (model scale) |
|---|---|
| *L* [m] | 3.826 |
| *LCG* [m] (from the front nose tip) | 1.76 |
| *LCB* [m] (from the front nose tip) | 1.76 |
| *VCG* [m] (from the keel) | 0.267 |
| *VCB* [m] (from the keel) | 0.289 |
| *m* [kg] | 703.976 |
| *Ixx* [kgm$^2$] (about the *CG*) | 30.849 |
| *Iyy* [kgm$^2$] (about the *CG*) | 665.448 |
| *Izz* [kgm$^2$] (about the *CG*) | 665.229 |

$$\delta_{P_{up}} = -\delta_r - \delta_s \quad (1)$$

$$\delta_{SB_{up}} = -\delta_r + \delta_s \quad (2)$$

$$\delta_{SB_{dn}} = \delta_r + \delta_s \quad (3)$$

$$\delta_{P_{dn}} = \delta_r - \delta_s \quad (4)$$

$$\delta_{sailplanes} = \delta_s \quad (5)$$

Thus, the effective rudder and sternplane angles were calculated as given in equations (6) and (7).

$$\delta_r = \frac{\delta_{P_{up}} - \delta_{P_{dn}} + \delta_{SB_{up}} - \delta_{SB_{dn}}}{4} \quad (6)$$

$$\delta_s = \frac{\delta_{P_{up}} + \delta_{P_{dn}} - \delta_{SB_{up}} - \delta_{SB_{dn}}}{4} \quad (7)$$

The autopilot and feedback controller utilised a Proportional-Derivative controller with associated proportional and differential control parameters for translations and rotations as used in the experiment [4].

The following three alternative approaches were used to predict the path of the submarine turning in the horizontal plane:

a) physical free running model [4];
b) fully coupled CFD simulation; and
c) Coefficient Based Model (CBM),

## 3. FREE RUNNING MODEL

The FRM tests were conducted by MARIN, and details are given in [4].

A 3.825m long model with particulars given in Table 1 was used. The tests were conducted in the Seakeeping and Manoeuvring Basin in Wageningen, Netherlands. The basin is filled with fresh water having a density of 1,000kg/m$^3$, however the data was given for a full scale submarine in sea water (1,025kg/m$^3$) which required scaling based on Froude scaling laws with a model scale factor of 18.348.

## 4. CFD MODEL

Details of the CFD Model are given in [3].

The CFD simulations utilised Unsteady Reynolds Averaged Navier Stokes (URANS) equations with turbulence modelled using the Elliptic Blending Reynolds Stress Model (EB RSM), coupled to a 6-DOF Rigid Body Dynamics (RBD) model [5]. The URANS equations solved the flow field for prediction of the vehicle's hydrodynamics. This was then applied to compute the vehicle's resulting motion using the RBD equations with linear and angular momentum solvers for translation and rotation of the vehicle defined as a rigid body with its mass properties as given in Table 1.

All simulations in this study were performed under transient conditions to investigate the behaviour of the vehicle in the time domain. Fresh water, with the same density as measured during the experiments was specified.

The CFD model involved the vehicle operating with movable control planes and a body force propeller (actuator disk). The latter is described using the propeller performance characteristics that were obtained from the captive self-propulsion simulation, which was carried out using CFD and employing the propeller geometry used in the experiment [4, 6].

The free running vehicle motion was applied to the complete computational mesh domain, including the vehicle and the background regions, while keeping the fluid stationary. This prevented degradation of the mesh quality that may otherwise have occurred as a result of mesh deformation to realise the vehicle's motion. The boundary conditions applied were velocity inlet with zero velocity at the inlet, pressure outlet at the sides, and no-slip wall on the vehicle. The boundary conditions led to reduction in the effects due to the proximity of the vehicle to the fluid domain boundaries, allowing a reduction in fluid domain size [5].

The dynamic deflection of the control planes was realised using the overset mesh [7] where the overset regions containing the ACSs were imposed on the background region and rotated at a rate commanded by the autopilot.

The actuator disk mimics the effects of a propeller by generating fluid acceleration based on the prescribed magnitude and distribution of fluid momentum sources. The axial and circumferential momentum sources were computed based on propeller performance characteristics. These were updated simultaneously with an observed inflow velocity at a plane placed forward of the disk origin by 10 % of the disk diameter. The distribution of the momentum sources was prescribed in accordance with the Goldstein optimum method [8] . The method prescribes the distribution of the axial (thrust/swept volume of the disk) and circumferential (torque/swept volume of the disk) momentum components over the radial direction of the actuator disk. For the actuator disk used in this study, the tangential velocities were neglected and hence the radial momentum component was zero [9]. The volume of the disk was determined to be equal to that of the propeller [10].

## 5. COEFFICIENT BASED MODEL

The CBM Submarine Simulation Model used was a 6-DOF, coefficient-based model developed by the Defence Science and Technology (DST) Group within a MATLAB/Simulink environment. The model's hydrodynamics representation was based on the equations developed by Gertler and Hagen, as discussed in [1]. They have been modified for the X-plane ACS configuration

The hydrodynamic coefficients for the BB2 were generated from a RANS CFD study. To simulate a controlled manoeuvre, the model implemented a proportional-derivative autopilot to set the angles of the submarine's control surfaces.

## 6. RESULTS

### 6.1 COMPARISON BETWEEN PREDICTIONS AND EXPERIMENTAL RESULTS

The results for a 20 degree effective rudder angle turn are presented in Figures 3 – 8. Unfortunately, due to difficulties with the autopilot, the sailplanes were fixed at an angle of zero degrees for both the FRM and the CFD model, whereas it was free to move for the CBM predictions. For the CBM simulations, the steady sailplane angle was about negative two degrees, which is very close to the balance angle required by the CBM for straight and level running.

As can be seen from Figure 3 the CFD model predicted the measured turning circle very well. However, the CBM predicted a much tighter turn with a smaller diameter. The steady state sway and yaw velocities predicted using the CBM were both greater than those measured, or predicted using the CFD model, as shown in Figures 4 and 5.

Figure 6 shows the pitch angle of the BB2 submarine as a function of time. As expected, the values of the measured and predicted pitch angles in steady state were all quite small, i.e. less than three degrees.

The angles of the ACSs as a function of time are presented in Figure 7. For a purely axisymmetric body with no asymmetry these angles should all be equal to 20 degrees.

However, the asymmetrical nature of the submarine geometry with the sail and casing generated a vertical downward force over the hull aft of the sail, shown by the green arrow in Figure 9 and discussed in [1]. The consequence of this is that the ACSs are required to generate a net upward force to compensate (open arrow in Figure 9). This is done by increasing the angle of the control surfaces generating an upward force, and decreasing the angle of those generating a downward force, as illustrated in Figure 10.

From Figure 7 it is seen that in the steady state the ACS angles required to maintain depth in the experiment were close to the effective rudder angle demand of 20 degrees. This implies that the downward force on the hull in the turn was not significant.

Equally, the required ACS angles predicted using the CFD model were also very close to 20 degrees.

However, the required ACS angles predicted using the CBM were not as close to 20 degrees. The ACSs generating upward forces ($\delta_{SB_{up}}$ and $\delta_{P_{dn}}$) were at significantly greater angles than 20 degrees, whereas the ACSs generating downward forces ($\delta_{SB_{dn}}$ and $\delta_{P_{up}}$) were at significantly lower angles than 20 degrees.

This implies that if the CBM was used to determine the required size of the ACSs to enable them to compensate for the downward force on the hull during a turn, it is likely that larger ACSs than required would be specified.

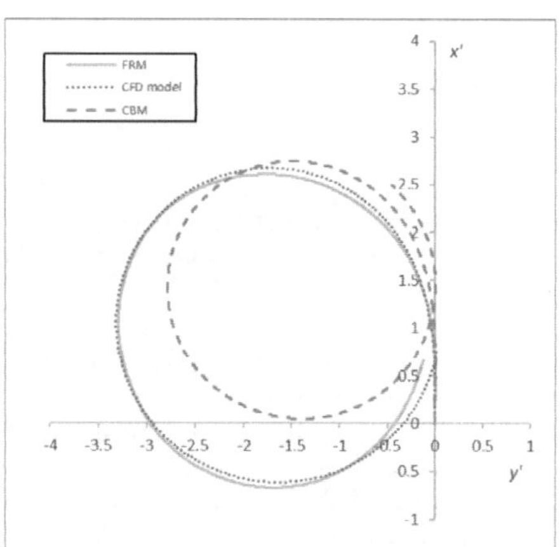

Figure 3. Turning circle of the BB2

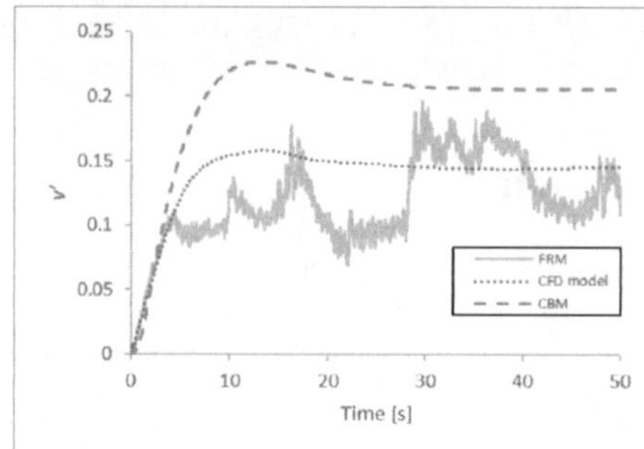

Figure 4. BB2 Sway velocity as a function of time

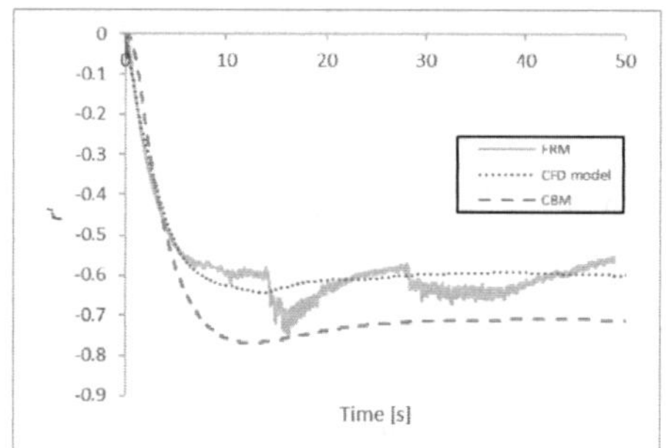

Figure 5. BB2 Yaw velocity as a function of time

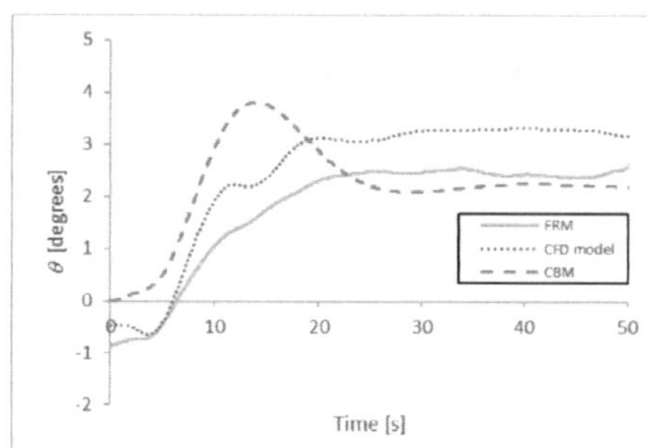

Figure 6. BB2 Pitch angle as a function of time

## 6.2 INFLUENCE OF CROSS COUPLING TERMS

The downward force on the hull when turning, as shown by the green arrow in Figure 9, is represented in the CBM by the following cross coupling terms: $Z'_{vv}, Z'_{rr}, M'_{vv}, M'_{vr}$ and $M'_{rr}$ [2]. To determine whether the values of these terms were the source for the ACSs to generate the much larger force in the CBM predictions than required in the experiments, the CBM was run with these terms set equal to zero.

The influence on the turning circle of setting these terms to zero was negligible, however the reduction in the vertical downward force on the submarine in the steady turn was significant. The outcome of this was that the predicted steady state pitch angle was zero when using the CBM method with the cross coupling coefficients set to zero, as seen in Figure 11.

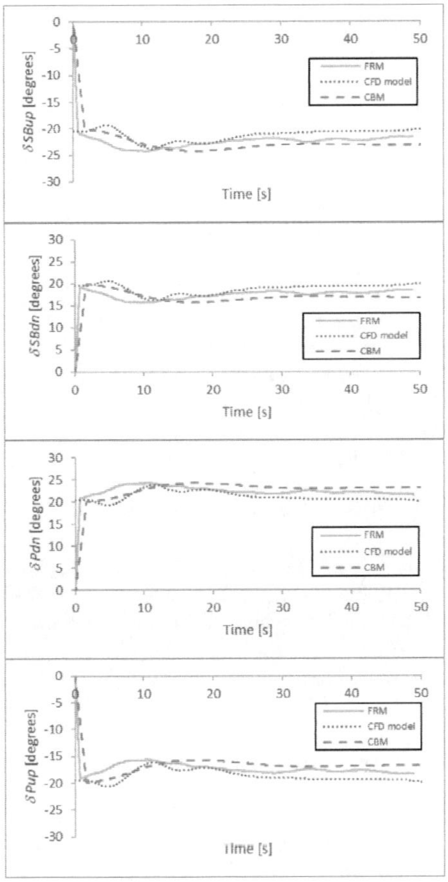


Figure 7. ACS angles of BB2 as functions of time

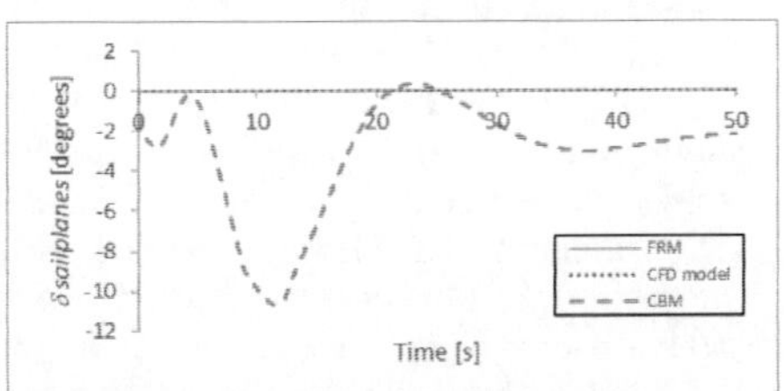


Figure 8. Sailplane angle of BB2 as function of time

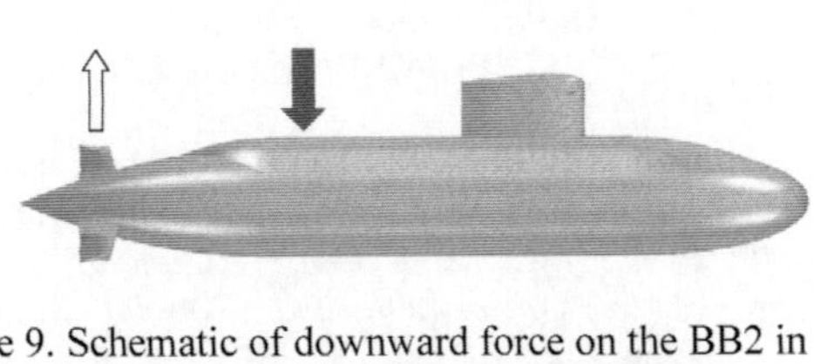

Figure 9. Schematic of downward force on the BB2 in turn due to asymmetry (green arrow), and upward force required from ACSs (open arrow)

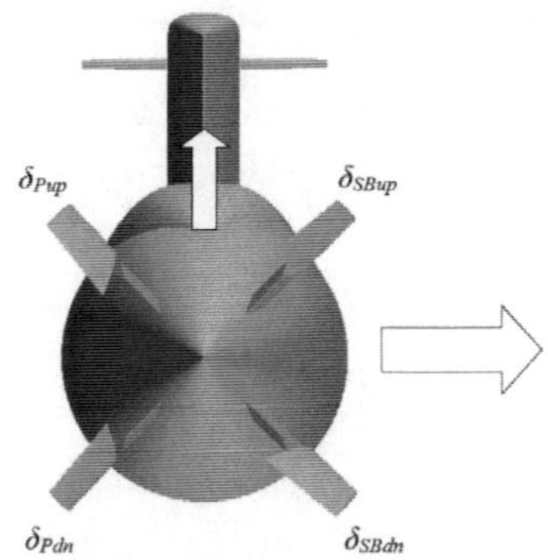


Figure 10. Operation of BB2 ACSs for turn to port

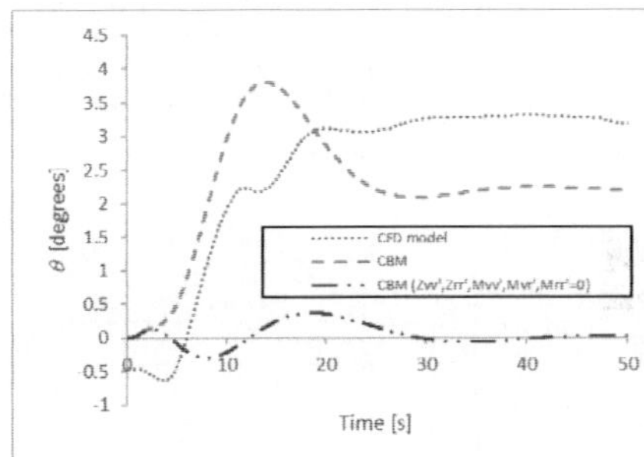


Figure 11. BB2 Pitch angle as a function of time

Consequently, the required ACS angles are much closer to the effective rudder angle of 20 degrees, as shown in Figure 12, and the sailplane angles are different, as shown in Figure 13.

## 7. DISCUSSION

An accurate determination of the actual angles required by the various ACSs to achieve the effective angle defined by equations 6 and 7 is significant in the design of the ACSs. If the ACSs are too small, and are not able to provide sufficient upward force, then they may not be able to compensate for the downward force on the submarine due to the out-of-plane loads, as discussed above.

This is of particular importance when maximum rate turns are being specified, as the planes which are required to attain a larger angle than the effective angle may be unable to do so, thereby limiting the rate of the turn.

Thus, it is important to be able to predict the influence of the out-of-plane loads on a submarine in the turn.

From the presented results it appears that the CBM, using coefficients obtained from steady state "captive" CFD simulations, over predicts the out-of-plane loads, and hence would lead to the design of larger ACSs than required.

At this stage it is not clear why this is the case, however this is the subject of an ongoing joint investigation by DST and the Australian Maritime College (AMC).

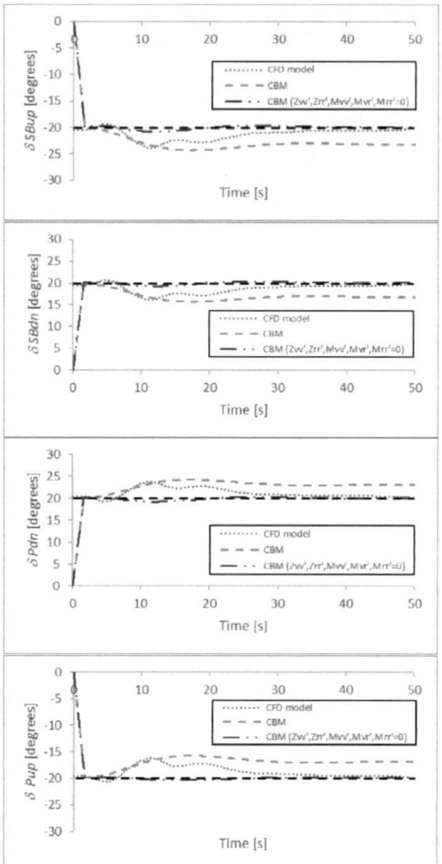

Figure 12. BB2 ACS angles as functions of time

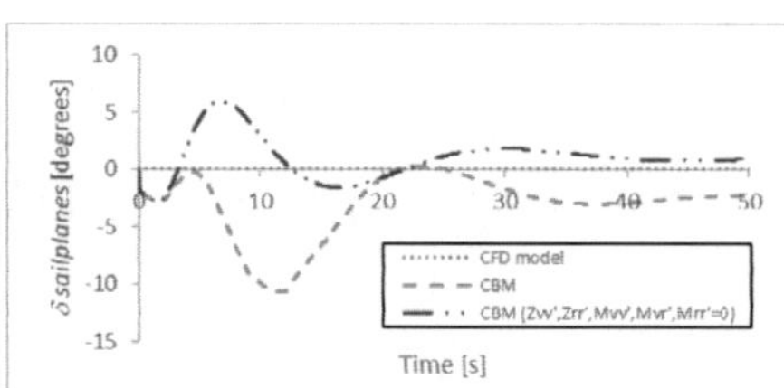

Figure 13. BB2 Sailplane angle as functions of time

## 8. CONCLUDING REMARKS

The presence of a downward force on a submarine hull aft of the sail results in the need for the ACSs to generate an upward force to compensate for this.

Based on the preliminary results presented in this paper, the upward force predicted using the CFD model is similar to that required on an FRM. However, the CBM appears to predict that a much larger ACS force is required. This is likely to be due to the way the downward force is estimated using the cross coupling terms in the equations within the CBM, as proposed by [2], and generally used in the submarine hydrodynamics community.

Further investigation is continuing into this as an understanding of the downward force on a submarine in a turn is necessary to ensure that the performance requirements for the ACSs are specified correctly.

## 9. ACKNOWLEDGEMENTS

The authors would like to thank Dr Anthony Fowler and Mr Alex Cameron for providing the CBM and assisting in its use. The authors are also grateful to Mr. Frans Quadvlieg from MARIN for providing the experimental data and the advice on its details.

## 10. REFERENCES

1. RENILSON, M. R., 'Submarine Hydrodynamics (Second Edition)', *Springer*, *ISBN 978-3-319-79056-5*, 2018.f
2. GERTLER, M., & HAGEN, G. R., 'Standard equations of motion for submarine simulation', *NSRDC Report 2510, David Taylor Naval Ship Research and Development Center*, 1967.
3. KIM, H., LEONG, Z. Q., RANMUTHUGALA, D., & CHIN, C., 'Six-DOF simulations of an underwater vehicle undergoing straight line and steady turning manoeuvres', *Ocean Engineering, 150, 102-112*, 2018.
4. OVERPELT, B., NIENHUIS, B., & Anderson, B., 'Free Running Manoeuvring Model Tests On a Modern Generic SSK Class Submarine (BB2)', *Pacific International Maritime Conference, Sydney*, 2015.
5. KIM, H., LEONG, Z. Q., RANMUTHUGALA, D., CHIN, C., & FORREST, A., 'Free running simulation of an Autonomous Underwater Vehicle undergoing a straight line manoeuvre via Computational Fluid Dynamics', *Pacific International Maritime Conference, Sydney*, 2015.
6. KIM, H., 'Six-DOF motion response and manoeuvring simulation of an underwater vehicle', *PhD Thesis, University of Tasmania/Australian Maritime College*, 2018.
7. CD adapco, 'STAR-CCM+ User Guide Version 10.02', 2015.
8. GOLDSTEIN, S., 'On the vortex theory of screw propellers', *Proceedings of the Royal Society of London. Series A, Containing Papers of a Mathematical and Physical Character, 123(792), 440-465*, 1929.
9. KIM, H., LEONG, Z. Q., RANMUTHUGALA, D., CHIN, C., & FORREST, A., 'Simulation and Validation of an AUV in Variable Accelerations', *International Journal of Offshore and Polar Engineering, 25(01), 35-44*, 2015.

10. PHILLIPS, A. B., TURNOCK, S. R., & FURLONG, M., 'Comparisons of CFD simulations and in-service data for the self propelled performance of an Autonomous Underwater Vehicle', *Symposium of Naval Hydrodynamics, Seoul*, 2008.

## 9. AUTHORS' BIOGRAPHY

**Howan Kim** holds the current position of Research Associate at Australian Maritime College (AMC), after completing his PhD at AMC. His current work is involved in CFD modelling of various experimental programmes to assess the hydrodynamic characteristics of underwater vehicles.

**Dev Ranmuthugala** is currently the Head, Hydroacoustics at the Defence Science and Technology (DST) Group, Department of Defence, Australia. Previous experience includes Professor in Maritime Engineering at the Australian Maritime College involved in hydrodynamics of underwater vehicles.

**Martin Renilson** holds the current position of Principal Hydrodynamicist at Pacific ESI. His previous experience includes as Technical Manager, Hydrodynamics, for QinetiQ in the UK. He is currently involved in assisting DST with its work on submarine hydrodynamics.

# HYDRODYNAMIC LOADS AND FLOW STRUCTURE ANALYSIS FOR A SURFACED SUBMARINE USING CAPTIVE MODEL EXPERIMENTS

**A Conway[1,2], C Kumar[1], A Rolls[2], A Fowler[1] and A Cameron[1,3]**, 1 - Defence Science and Technology Group, AUS
2 – Australian Maritime College, AUS 3 – QinetiQ Australia, AUS

## SUMMARY

Submarines are required to operate on the surface for significant periods of time, for example during transit in shallow waters. However, as they are designed primarily for submerged operations, their surfaced sea-keeping behaviour may be poor. Furthermore, the wave making resistance of the vehicle increases in the surfaced state, substantially impacting on its powering requirements. The bow wave encountered when operating in a surfaced state applies a downward heave force and a bow down pitching moment, impacting the vessel's sea-keeping performance.

To provide insight into these effects, a captive model experimental program was undertaken at the Australian Maritime College Towing Tank facility to generate data at model scale for a 1.69m BB2 generic submarine geometry operating on the surface in calm water. The data generated includes the hydrodynamic loads and detailed, time-averaged flow structures on the surface, which will be used to study the behaviour of the vehicle and validate numerical tools.

## NOMENCLATURE

| | |
|---|---|
| $\beta$ | Angle of drift (deg) |
| $\rho$ | Density of water (kg m$^{-3}$) |
| $C_x$, $C_y$, $C_z$ | Coefficient of surge, sway and heave |
| $C_m$, $C_n$ | Coefficient of pitch and yaw |
| $Fr$ | Froude number |
| $g$ | Acceleration due to gravity (m s$^{-2}$) |
| $L$ | Boat length (m) |
| $Re$ | Reynolds number |
| $U$ | Forward velocity (m s$^{-1}$) |
| $X$, $Y$, $Z$ | Surge, sway and heave force (N) |
| $M$, $N$ | Pitch and yaw moments (Nm) |

## 1. INTRODUCTION

Naval submarines have an operational requirement to spend extended periods of time on the surface. While surfaced, the vessel must operate at a relatively slow speed, which has a negative impact on the vessel's manoeuvrability and control. The vessel is also subjected to various hazards such as beam winds and high sea states, which can impact its stability [1]. As current generations of submarines prioritise submerged performance over sea-keeping, which is generally poor, understanding these parameters becomes critical [1], especially with respect to control. Furthermore, the wave making resistance of the vehicle increases in the surfaced state, substantially impacting on its manoeuvring coefficients. The bow wave encountered when operating in a surfaced state applies a downward heave force and a bow down pitching moment, impacting the vessel's sea-keeping performance. These interactions are complex, and a combination of numerical simulations and experimental programs is often employed to provide insight.

Computational fluid dynamics (CFD) simulations employing Reynolds-averaged Navier-Stokes (RANS) equations is one method used to estimate these behaviours. However, these methods are reliant on factors such as turbulence models, grid resolution or meshing techniques, each of which can predict significantly different viscous and pressure forces. These factors can substantially influence the manoeuvring coefficients estimated. In order to ensure the most appropriate approach is chosen, these simulations require a robust experimental dataset for validation purposes. This report presents the first in a series of surfaced captive model experiments to generate a dataset which can be used to develop the required coefficients for manoeuvring models.

The experiment was undertaken on a 1.69m long surfaced submarine model in calm water, with the geometry being based on the BB2 generic diesel-electric submarine. The experiments consisted of two stages. Firstly, a new calibration technique was developed to produce high-quality loads measurements. Secondly, a series of straight line and static angle of drift measurements were undertaken at two Froude numbers. Force measurements were complemented with surface flow field visualisation of bow wave height and flow structures in the bow region. Assessment of the experimental uncertainty is provided.

This paper demonstrates the growing capability within the Defence Science and Technology (DST) Group and the Australian Maritime College (AMC) to measure and predict hydrodynamic loads and free surface interactions at model-scale Froude numbers.

## 2. MODEL DETAILS

The submarine model used in these tests, herein referred to as BB2, is based on the geometry proposed by [2] and is shown in Figure 1. The geometry was designed for the purpose of research studies and has a full-scale length of

 

70.20m, while the scaled model used has an overall length ($L$) of 1.69m. The model is axisymmetric for the first $0.07L$ from the nose tip, with the nose derived from a NACA-0018 [3] forebody. It has a casing, sail (NACA-0022) and X-form aft control surfaces (ACS).

The model consists of three detachable, waterproof sections - the nose cone, central hull and tail cone. The model was attached to the carriage via the sail, as shown in Figure 2. The horizontal planar motion mechanism (HPMM) was used to hold the model rigidly in place and set the model's static drift angles.

The model skin is attached to the two strain gauges located in the central hull, as shown in Figure 2. The interior has space for attaching ballasting weights to enable the model to be appropriately trimmed, as per ITTC Recommended Procedure [4]. Waterline markings are used on the exterior skin to ensure the model operates at the BB2's standard surfaced operating condition. The full-scale displacement is approximately 3,788 tonnes (51.71kg model scale).

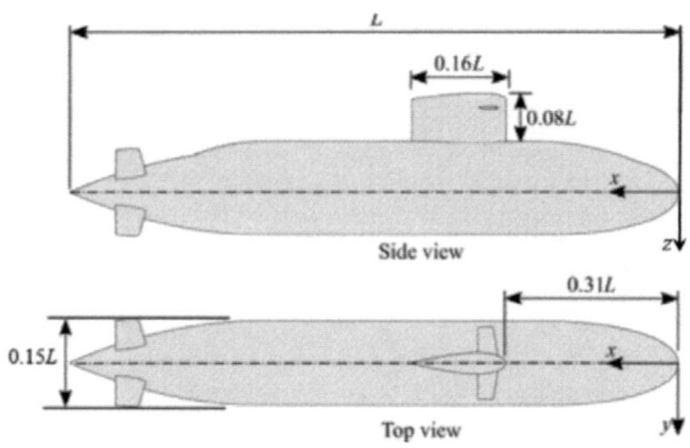


Figure 1: Schematic of the BB2 submarine geometry, where $L$ is the model length of 1.69 m.

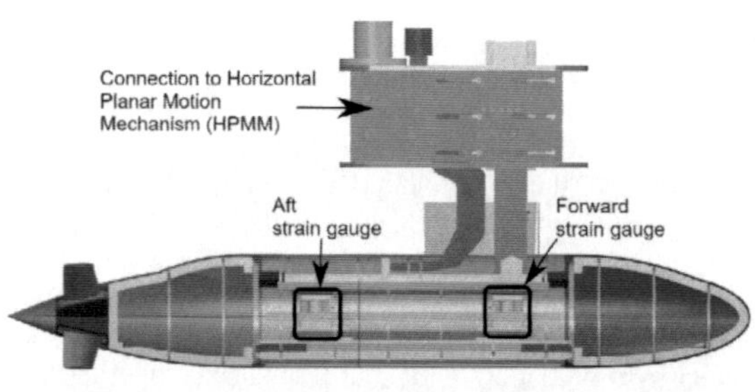


Figure 2: Cross-section of the AMC BB2 captive model showing the location of the AMTI strain gauges and the attachment of the model to the horizontal planar motion mechanism (HPMM) via the sail.

## 3.1 METROLOGY

In order to quantify the model's conformity to its computer aided design (CAD) design file, a 3D scan of the model was undertaken using a 3D system laser scanner with a reported accuracy of ±0.25mm. The system estimated that the model varied from the design file by an average of 0.28mm, as shown in Figure 3. While the overall accuracy of the model is close to the laser scanner's measurement uncertainty, certain regions of the model varied more substantially. Asymmetry was noted between the model's starboard and port side which could have contributed to the loads measurement uncertainty, as further discussed below.

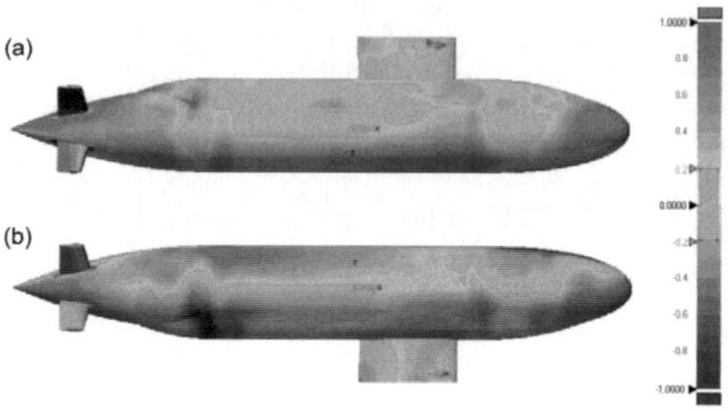


Figure 3: Laser scan of the (a) starboard (b) and port side of the BB2 surfaced model. All dimensions in mm.

## 3.1 TURBULENCE STIMULATION

The model size and forward velocities used during this test program resulted in low Reynolds numbers ($Re < 1.5 \times 10^6$). In order to ensure a consistent turbulent flow regime across the length of the model, artificial turbulence generators were used to ensure the boundary layer characteristics were representative of a full-scale vessel. This consisted of a regular saw-tooth pattern Hama trip of a constant thickness (0.52mm) installed at 5% of the model length from the bow, as shown in Figure 4. The Hama trip height and tooth pattern were based on the work of [5] and are shown to provide a consistent turbulent regime across the length of the model whilst minimising parasitic drag.

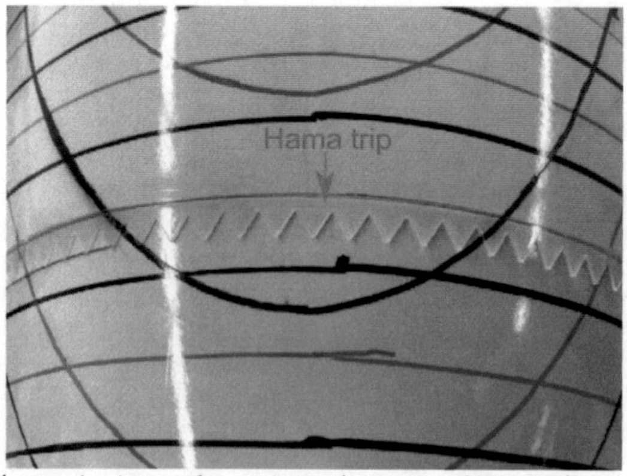


Figure 4: A regular saw-tooth pattern Hama trip was installed at 5% of the model length from the bow to ensure the boundary layer characteristics were representative of a full-scale vessel.

 

As the sail was above the surface, no tripping was required. The ACS were not tripped due to challenges in correctly implementing a trip height in the Reynolds number range encountered. As the contribution of the ACS to the global loads is minimal [6], the lack of tripping is unlikely to have an undue influence on overall results.

## 4. LOADS MEASUREMENT

The model's hydrodynamic loads were measured via two internally mounted AMTI MC3A-100 6-axis strain gauge sensors that were rigidly connected to the model's central hull via a strong-back, as shown in Figure 2.

Previous experimental programs at the AMC performed the strain gauge calibration outside the model and calibrated each principal load axis individually. After calibration, the strain gauges were installed in the model using a hinge and slider method and the outputs 'summed' as described by [7]. This technique did not account for interactions between the strain gauge channels upon application of a mass, referred to as crosstalk. This has the potential to increase the measurement uncertainty.

The current design, based on the work of Kumar [5], removed the 'slider and hinge' and replaced it by rigidly connecting the two strain gauges inside the model via a precisely machined strong-back. The fully assembled model was installed on a custom designed rig, as shown in Figure 5. The entire model was subsequently calibrated as an integrated unit.

The rig was composed of two frames. The first frame (red) was constructed using hollow steel sections with a thick steel plate at the top for rigidly connecting the model onto the frame via the sail. The frame, and thus model, were levelled in roll and pitch. The second frame (blue) was constructed using Thorlabs XT95 precision rails and aligned various calibration devices to the model.

Figure 5: Custom calibration rig designed for the BB2 surfaced model. The red part was used to rigidly support and level the model. The blue part was composed of Thorlabs XT95 precision rails for the mounting and alignment of calibration devices.

Calibration was achieved by applying calibrated weights to each load direction at specified tapped hard points on the model, as shown in Figure 6. Each hard point represented one load condition and resulted in either a 'pure' force, when loaded orthogonal to the model origin, or a force and moment combination when offset from the model origin. The locations of the hard-points and resulting forces and moments are summarised in Table 1.

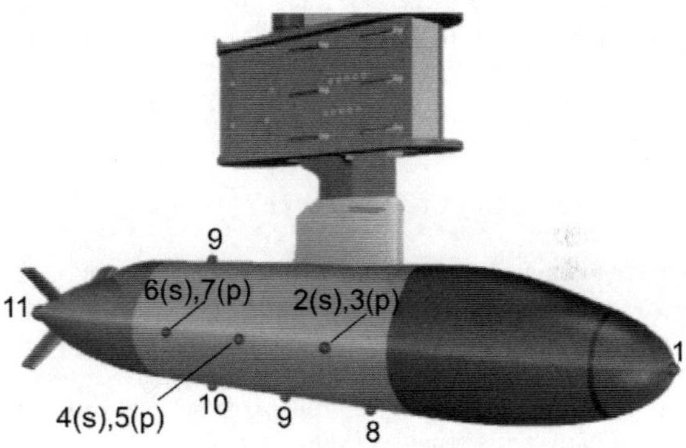


Figure 6: Positions of tapped hard points used for model force and moment calibration. Note the side taps were present on both the starboard (s) and port (p) sides.

Table 1: Location of tapped hard-points on model and resulting forces and moments.

| Load Point | Model Location | Component |
|---|---|---|
| 1 | Nose centreline | -X |
| 2, 3 | Starboard/Port forward | ±Y & ±N |
| 4, 5 | Starboard/Port centre | ±Y |
| 6, 7 | Starboard/Port aft | ±Y & ±N |
| 8 | Keel forward | +Z & -M |
| 9 | Keel centre | +Z |
| 10 | Keel aft | +Z & +M |
| 11 | Stern centreline | +X |

The calibration matrix was determined by the application of 11 load cases by

$$[C] = [F] \cdot [V - V_0]^T$$

Where $[C]$ is a 6×6 calibration matrix, $[F]$ is a 6×1 matrix of input weights representing forces and moments and $[V - V_0]$ is a 6×1 matrix that combines the forward and aft strain gauge voltage output ($V$) minus a voltage offset ($V_o$) in the surge, heave and sway channels of each strain gauge. $T$ is the matrix transpose function. The voltage offset, captured for 5 seconds, removed factors such as data acquisition electrical noise and model mass.

Using the strain gauge data recorded during the experiment, the hydrodynamic loads were subsequently determined by

$$[F] = [C] \cdot [V - V_0]$$

All load directions, with the exception of heave, were applied using air bearings to reduce frictional losses. The loads were applied and removed in increments up to the peak load, as estimated by preliminary CFD simulations with a safety margin of ±50%. As is required for a 6×6 calibration matrix, the strain gauge response was linear for each load point.

The calibration matrix accuracy was estimated by applying a series of multi-axis load checks. This involved loading multiple force and moment channels simultaneously, recording the output and comparing it to the applied loads.

### 4.1 RANDOM UNCERTAINTY

The random uncertainty was estimated by repeating a single configuration 11 times over the test program duration (Fr=0.16, β=0°). The effect of speed and angle of drift on random uncertainty can be investigated in future experimental programs. The repeat runs were filtered using Chauvenet's criterion [8] to remove outliers.

The random uncertainty was combined with the calibration uncertainty, discussed above, to determine the overall uncertainty of the experiment.

## 5. TEST MATRIX

Two speeds and a range of static drift angles were selected in order to simulate the realistic operating conditions of a submarine whilst in the surfaced condition. A summary of the experimental test program is presented in Table 2. All speeds are presented as non-dimensional Froude numbers based on the length of the model and defined as:

$$Fr = \frac{U}{\sqrt{gL}}$$

where $U$ is the forward velocity, $g$ is the acceleration due to gravity and $L$ is the length of the model.

Table 2: Froude numbers and drift angles tested during the experimental program.

| Froude number | Angle of drift (°) | Increments (°) |
|---|---|---|
| 0.16 | $-8° \leq \beta \leq +8°$ | 2° |
| 0.22 | $-8° \leq \beta \leq +8°$ | 2° |

## 6. FLOW VISUALISATION

The flow structures around the bow of the model, in particular the bow wave height, were captured using two Canon EOS 1D X Mark II cameras installed with either a Canon EF 50mm f/1.8 STM or Canon 24mm f/2.8 STM lens. As shown in Figure 7, camera one was positioned on the starboard side of the model, capturing the vertical bow wave, whilst camera two was positioned above the model, capturing the longitudinal position of the bow wave as well as the start of the Kelvin wake.

Both cameras were triggered from the data acquisition software that also recorded the strain gauge outputs, with the run being captured from start to finish at 10 frames per second. This ensured that the images were correctly synchronised with the force and moment data collected.

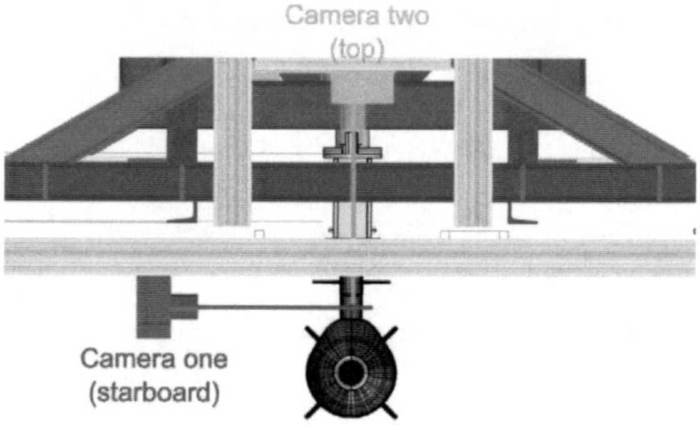


Figure 7: Front view of the towing tank (TT) carriage setup with the surfaced BB2 model and camera one (red) and two (green).

## 7. RESULTS AND DISCUSSION

### 7.1 CALIBRATION

As mentioned in Section 4, the loads measurements calibration uncertainty was estimated through the application of multi-axis load checks. A total of 7 load cases were applied, resulting in various combinations of forces and moments which are summarised in Table 2.

Table 2: Summary of complex loads applied to determine calibration uncertainty.

| Complex load case | Applied loads | | | | |
|---|---|---|---|---|---|
| | X (N) | Y (N) | Z (N) | M (Nm) | N (Nm) |
| 1 | 0.00 | 0.00 | 0.00 | 0.00 | 0.00 |
| 2 | 4.92 | 0.00 | 0.00 | 0.00 | 0.00 |
| 3 | 4.92 | -9.83 | 0.00 | 0.00 | -2.15 |
| 4 | 4.92 | -9.83 | 19.63 | -4.28 | -2.15 |
| 5 | 0.00 | -9.83 | 19.63 | -4.28 | -2.15 |
| 6 | 0.00 | 0.00 | 19.63 | -4.28 | 0.00 |
| 7 | 0.00 | 0.00 | 0.00 | 0.00 | 0.00 |

Table 3: Percentage error recorded when comparing applied loads to recorded loads during the complex loads uncertainty analysis.

| Complex load case | Error | | | | |
|---|---|---|---|---|---|
| | X (%) | Y (%) | Z (%) | M (%) | N (%) |
| 1 | 0.00 | 0.01 | 0.02 | 0.02 | 0.02 |
| 2 | 0.46 | 0.13 | 0.12 | 0.12 | 1.06 |
| 3 | 0.05 | 0.80 | 0.37 | 0.04 | 0.28 |
| 4 | 0.08 | 0.67 | 0.12 | 0.20 | 0.42 |
| 5 | 0.52 | 0.72 | 0.06 | 0.03 | 0.66 |
| 6 | 0.03 | 0.36 | 0.14 | 0.12 | 0.30 |
| 7 | 0.03 | 0.31 | 0.09 | 0.04 | 0.21 |

As can be seen in Table 3, the maximum uncertainty due to calibration is less than 1%. The accuracy in applying the forces and moments on the model during the calibration phase has resulted in a reduction in calibration uncertainty, and thus, total experimental uncertainty. This has resulted in a significant increase in the quality of the data.

## 7.2 FORCE AND MOMENT DATA

The force and moment data collected during this experimental program uses the body-fixed coordinate system and all data is non-dimensionalised with respect to length. The equations used are:

$$C_x = \frac{X}{\frac{1}{2}\rho U^2 L^2}$$

$$C_y = \frac{Y}{\frac{1}{2}\rho U^2 L^2}$$

$$C_z = \frac{Z}{\frac{1}{2}\rho U^2 L^2}$$

$$C_m = \frac{M}{\frac{1}{2}\rho U^2 L^3}$$

$$C_n = \frac{N}{\frac{1}{2}\rho U^2 L^3}$$

where $C_x$, $C_y$, $C_z$, $C_m$ and $C_n$ are the coefficients of drag, sway, heave, pitch and yaw respectively. $X$, $Y$, $Z$ represent the surge, sway, heave forces and $M$ and $N$ are the pitch and yaw moments. ρ is the density of water, $U$ is the forward velocity of the model and $L$ is the overall length of the model.

Figures 8 – 10 show the force and moment data plotted against the angle of drift for both Froude numbers tested.

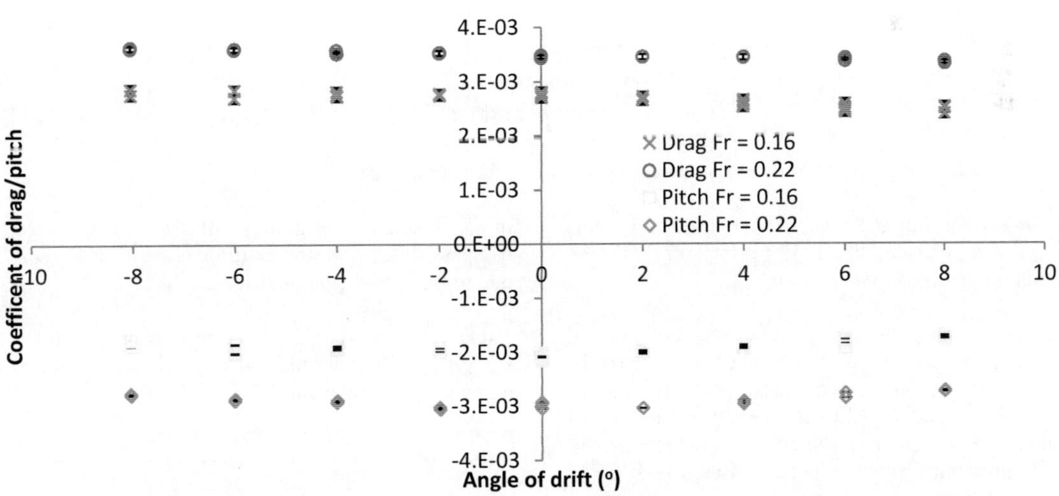


Figure 8: Coefficient of drag ($C_x$) and coefficient of pitch ($C_m$) plotted against angle of drift for Froude numbers 0.16 and 0.22.

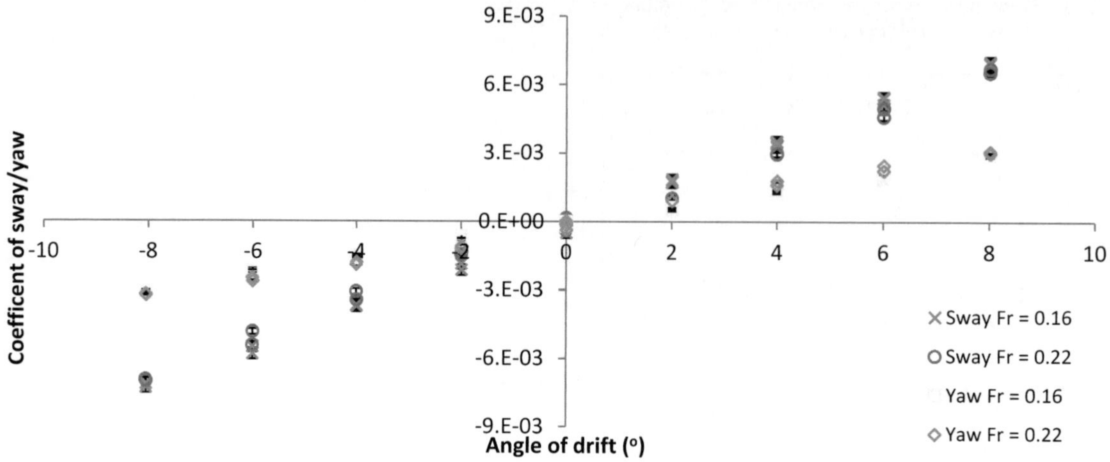


Figure 9: Coefficient of sway ($C_y$) and coefficient of yaw ($C_n$) plotted against angle of drift for Froude numbers 0.16 and 0.22.

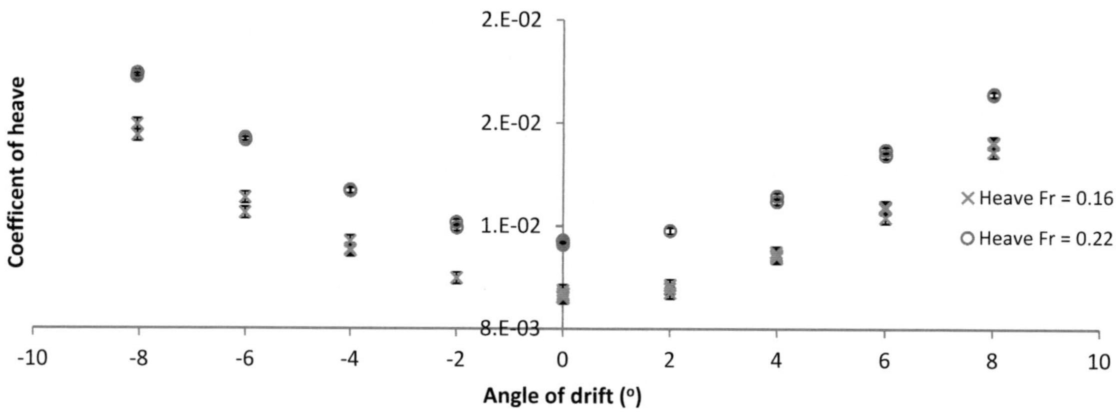


Figure 10: Coefficient of heave plotted against angle of drift for Froude numbers 0.16 and 0.22.

As can be seen in Figures 8 - 10 above, the coefficients show excellent repeatability (<1% error) for both the forces and moments.

There are two notable trends within the force and moment data. Firstly, we can see clearly defined linear trends for the sway and yaw coefficients and quadratic trends for the pitch and heave coefficients. It should be noted, as mentioned earlier, that the angle of drift is with respect to flow direction and forces are presented as a body-fixed coordinate system. This results in the sway and yaw coefficients appearing to be Froude number independent. If the data was presented as flow-fixed, the sway and yaw coefficients would not be Froude number independent.

Secondly, there is an asymmetry in the data when comparing negative and positive drift angle values which is presented in Table 4. The calibration was tested in both directions to ensure the calibration was not contributing to the asymmetry in the results. It is hypothesised that the variations in model's geometry, as shown in Figure 3, likely contributed to the asymmetry in the force and moment data. Future studies will undertake CFD analysis, using 3D laser scan mesh, to determine loads sensitivity due to the model's geometric variation.

Table 4. Percentage difference between port and starboard surge coefficient at angle of drift ($\beta$) sweeps for Froude number (Fr) of 0.16 and 0.22

| Fr | $\beta$° | *% difference $C_x$* |
|---|---|---|
| 0.16 | 2 | 3.6 |
| | 4 | 7.1 |
| | 6 | 10.3 |
| | 8 | 13.5 |
| 0.22 | 2 | 2.0 |
| | 4 | 2.8 |
| | 6 | 5.8 |
| | 8 | 8.2 |

### 7.3 FLOW VISUALISATION RESULTS

While flow visualisation does not directly contribute to the calculation of manoeuvring coefficients, it can be a critical tool in the validation process for numerical simulations. With an increase in accuracy for numerical validation, the coefficients obtained from numerical simulations should benefit from a reduction in overall error and uncertainty. The results from the straight ahead cases of the flow visualisation are presented in Figure 11.

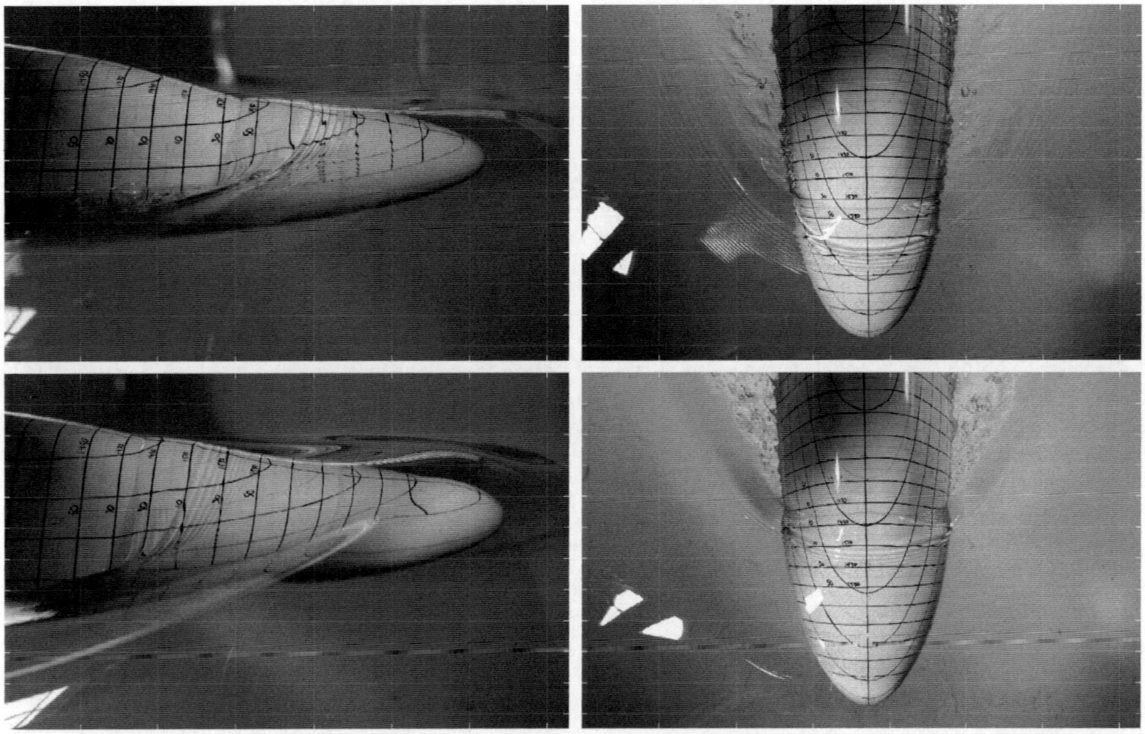

Figure 11: Side (left) and top (right) views of BB2 model bow region showing the flow visualisation for Froude number 0.16 (top) and 0.22 (bottom).

Both the side and top views of the bow region clearly show the bow wave run up, wave trough on each side and the development of the Kelvin wake. Using the clearly marked grid lines drawn on the model, accurate measurements can be recorded to further understand the hydrodynamic flow structures on the body.

## 7. CONCLUSIONS

This paper presents the results of a captive model experiment undertaken on a surfaced submarine at the towing tank facility at the Australian Maritime College using a 1.69m generic (BB2) submarine model. The model was tested in calm water and under the standard surfaced operating condition of a typical 70m submarine. The experiment was performed at two Froude numbers and across a range of angles of drift. Both starboard and port side angle-of-drift sweeps were undertaken to confirm experimental symmetry. The steady state hydrodynamic loads and surface flow visualisation were captured to validate numerical simulations. These simulations, in turn, will be used to generate coefficients for manoeuvring models.

A novel calibration process that replicated the in-tank setup was used for the first time. The fully assembled model was installed on a custom designed rig and a series of load cases were applied to generate a 6×6 calibration matrix. This process ensured potential sources of error, such as strain-gauge crosstalk, were accounted for. The resulting calibration was shown to be robust, with the calibration uncertainty being less than 1% across all forces and moments.

The experimental results were shown to be repeatable for both forces and moments at all angles of drifts. Well defined trends for sway, yaw, heave and pitch were shown, however, the surge coefficients showed a higher level of asymmetry that was outside the bounds of the measurement uncertainty. It is hypothesised that variations in the model's geometry, as a result of the manufacturing process, resulted in this asymmetry. Future

CFD simulations, utilising the model's laser scan mesh, will seek to quantify the impact of such geometric deviations on force and moment measurements. A CFD analysis could also quantify the sensitivity of manoeuvring coefficients as a result of the model asymmetry.

This experiment also undertook surface flow visualisation, which generated high quality images capturing the flow structures. The starboard camera proved to be effective in estimating the bow wave run-up on the model and the top camera was effective in capturing the surface wave patterns. Combined with the force and moment data, this preliminary dataset provides two significant points for future numerical validation studies.

## 8. ACKNOWLEDGEMENTS

The authors would like to acknowledge the team at the Australian Maritime College for their hard work and continued pursuit of improvements in model construction and experimental testing procedures. In particular, the authors thank Liam Honeychurch for the model construction, Kye Curwen and Toby Davis for technical support and Jonathan Duffy for invaluable input and feedback during all stages of the experimental program.

## 9. REFERENCES

1. LETTER, B., RANMUTHUGALA, D., DUFFY, J., TSO, Y., 'Numerical and Experimental Analysis of a Surfaced Submarine in Beam Seas', *International Maritime Conference*, 2010.

2. OVERPELT, B., NIENHUIS, B., ANDERSON, B., 'Free Running Manoeuvring Model Tests On A Modern Generic SSK Class Submarine (BB2)', *PACIFIC 2015*, 2015.

3. JACOBS, N., WARD, K., PINKERTON, R., 'The Characteristics of 78 Related Airfoil Sections from Tests in the Variable-Density Wind Tunnel', *NASA*, 1933.

4. ITTC, 'Recommended Procedures and Guidelines, Model Manufacture Ship Models, 7.5-01-01-01, Revision 01', *International Towing Tank Conference*, 2002.

5. KUMAR, C. et al., 'Boundary layer tripping study on a Generic Submarine Model in Water', *21st Australasian Fluid Mechanics Conference*, 2018.

6. SEIL, G., POOK, D., NGUYEN, M., LEONG, Z., 'The Influence of Appendages and their Stall on Submarine Hydrodynamic Loads', *PACIFIC 2017*, 2017.

7. WILSON-HAFFENDEN, S. J., 'An Investigation into the Wave Making Resistance of a Submarine Travelling Below the Free Surface', *Australian Maritime College: Launceston, Tasmania*, 2009.

8. LIN, L., SHERMAN, P., 'Cleaning Data the Chauvenet Way', *SESUG proceedings*, 2018.

## 9. AUTHORS BIOGRAPHY

**Alexander Conway** holds the current position of research fellow at the Australian Maritime College and is positioned at Defence Science and Technology Group at Fishermans Bend, Australia. He is responsible for both numerical and experimental modelling of underwater vehicles in order to determine the forces required to calculate the manoeuvring coefficients.

**Chetan Kumar** is a Maritime Experimental Specialist with the Defence Science and Technology Group in Fishermans Bend, Australia. His interests include loads and wake measurements of maritime platforms.

**Adam Rolls** holds the current position of research engineer at the Australian Maritime College. He is responsible for experimental research focussing on collaborations with external partners include defence and commercial vessel designers. His previous experience includes experimental investigations into the behaviour of surfaced vessels.

**Anthony Fowler** is a Maritime Platform Dynamics and Control Specialist with the Defence Science and Technology Group in Fishermans Bend, Australia. His interests include the modelling and simulation of maritime platforms and the development of associated control systems.

**Alexander Cameron** holds the current position of Engineer at QinetiQ and is positioned at Defence Science and Technology Group at Fishermans Bend, Australia. He is responsible for the development and testing of simulation models for the purpose of investigating manoeuvring, control and dynamic performance of underwater vehicles.

 

# THE DESIGN AND HYDRODYNAMIC ASSESSMENT OF A SUBMARINE CONCEPT WITH OFF-AXIS PROPULSION

**P Crossland,** QinetiQ Ltd, UK
**R Crocker,** BAE Systems–Maritime Submarines, UK
**S Machin** Submarine Delivery Agency, UK

## SUMMARY

The UK's approach to submarine design is largely evolutionary, with future submarine hull form designs likely to remain relatively conventional. However, changing naval strategy means that there is still sufficient justification to perform investigations into novel hull forms, which have the potential to offer easier integration of required capabilities or provide the ability to incorporate novel technologies. This paper describes a challenging design and hydrodynamic assessment exercise, in which BAE Systems–Maritime Submarines were tasked, by the Submarine Delivery Agency, to develop a concept design for evaluation by QinetiQ using their computation fluid dynamics and experimental assessment techniques. This work also provides the opportunity to de-risk novel design features to inform future underwater platform programmes and by undertaking a full manoeuvring and control evaluation of this design, the traditional testing techniques were challenged and developed.

## NOMENCLATURE

| | |
|---|---|
| BG | Distance between the vertical centre of buoyancy and gravity (m) |
| $G_h$ | Horizontal plane stability index (-) |
| $G_v$ | Vertical plane stability index (-) |
| L | Length of submarine (m) |
| n′ | Propulsion state (-) |
| U | Forward speed ($ms^{-1}$) |
| X′ | Non-dimensional axial force |
| $X_n$ | Propulsion lateral lever arm (Nm) |
| Z′ | Non-dimensional heave force |
| M′ | Non-dimensional pitch moment |
| $\rho$ | Density of water ($kg\ m^{-3}$) |

## 1. INTRODUCTION

One aspect of submarine design that is non-negotiable is safety and so there will always be a requirement to manage the safety of a submarine throughout its lifecycle. This requirement to manage the safety of a submarine, whilst at sea, necessitates a number of key factors to be understood; one of those factors is the ability to predict the manoeuvring behaviour of the submarine throughout the design process.

Any predictive capability should be able to safely explore the submarine manoeuvring envelope during the early design phases, to ensure unsuitable designs are not taken through to detailed design or build, and to ultimately provide safety guidance for operations at sea. The approach adopted in the UK has been developed from knowledge of hydrodynamics gained over many years of assessing submarines with a conventional hull form design. The outcome is to provide a validated toolset and processes that provide the evidence for assuring submarine safety, [1], which has benefitted:

- Future projects groups investigating submarine design at a concept level;
- Acquisition design groups carrying out more detailed project definition assessments; and
- In-service support to set safety and operational constraints for the Royal Navy (RN) fleet.

The last submarine type operated by the RN, that had twin propulsion, was the Oberon Class. These submarines, designed for Cold War operations, were finally retired from RN service in the 1990s. These diesel-electric submarines would operate at slow speed when submerged meaning the safety assurance processes, used for manoeuvring of submarines of today, would have been less relevant.

Since the Oberon Class, submarines designs have generally become an evolution of previous classes. Whilst future submarine hull form designs are likely to remain relatively conventional, changing naval strategy means that there may be sufficient justification to perform investigations into novel hull forms; these hull forms may have the potential to offer easier integration of required capabilities or provide the ability to incorporate novel technologies. However, if a submarine hull form design changes radically, the tools and processes for evaluating performance need to be re-evaluated and developed, in advance, to ensure the capability to assess that design is maintained.

To support the UK MOD in assuring the safety of their submarines, there is a requirement on QinetiQ to maintain a capability to evaluate the Manoeuvring and Control (M&C) performance of a submarine through the Maritime Strategic Capability Agreement for Submarine Hydromechanics. This capability includes maintaining the ability to predict and model (both physically and computationally) the hydrodynamic aspects and manoeuvring performance of existing and new submarine designs, including appendage design and, in particular, their ability to recover from emergency scenarios.

In terms of a programme of work, this means that QinetiQ undertakes a nominal three-year rolling cycle of model build and testing that includes:

- Captive model build and testing in the Ship Tank and the Rotating Arm
- Free-running model build and testing in the Ocean Basin and deep-water facilities
- Complimentary CFD

This three year rolling cycle can be traced back to the support QinetiQ provided to BAE in the development of the Astute Class hull form and has been repeated since, see [2] for example, and more recently in support of Dreadnought. To further exercise this capability, QinetiQ was provided, by UK MOD, with a challenging design that would exercise and extend the M&C capability over the three year rolling cycle. The requirement to test this concept design was driven by the desire to explore potential concept features and overall architecture that could be of benefit to future platforms; concept features include a novel propulsion off-axis propulsion system. BAE were tasked by the UK MOD to develop a concept design and provide the design to QinetiQ for evaluation using their standard approaches including computation fluid dynamics and physical model testing.

## 2. CONCEPT DESIGN

The UK has developed and applied M&C testing techniques to evaluating submarine configurations that have been introduced into service with the RN since the 1960's. QinetiQ Haslar has built and tested physical models of these designs in facilities such as the Ship Tank with the Planar Motion Mechanism (PMM), the Rotating Arm and the Ocean Basin and in a deep-water facility. These submarine designs have 'traditional' hull and appendage configurations comprising:

- A body of revolution with length/diameter ratio of ranging from 8-12 depending on platform type.
- Top-casing.
- Bridge fin.
- Cruciform stern appendages.
- Bow hydroplanes.
- Propulsor on the axis line at the tail cone aft of the stern appendages.

The primary established techniques for assessment are currently:

- Measurements on physical models (typically models are 4.5 to 5.5m long) tested in the Haslar facilities.
- A range of mathematical modelling and computer simulation techniques.

The application of these established techniques to significantly different design concepts or different operational modes has necessitated modifications to be made to the suite of tools and techniques. For example, see [2], which assessed a hull form with an X-plane stern configuration. This configuration led to the development of a new capability to evaluate X-planes and to be able to develop SOEs for such designs.

There were number of constraints agreed at the start of the concept design study, namely that;

- The basis for the design will be a single, typical, pressure hull.
- The model will need to operate in existing QinetiQ facilities and interface with existing infrastructure;
- The free-running model will be compatible with the current QinetiQ capability;
- The operational profile is consistent with that expected of the current in-service submarine fleet.

In addition to these constraints, there were a number of individual requirements imposed on the general arrangement, at concept level; that the design represented what was considered a credible design with the principal submarine systems distributed in a rational manner and the design had to be hydrostatically balanced, with a credible BG, when submerged.

The focus of the design was to provide a hull form with an enhanced launch and recovery capability of various payload configurations and a novel off-axis propulsion system was identified as a technology enabling strategy that could facilitate an enhanced payload design and integration solution.

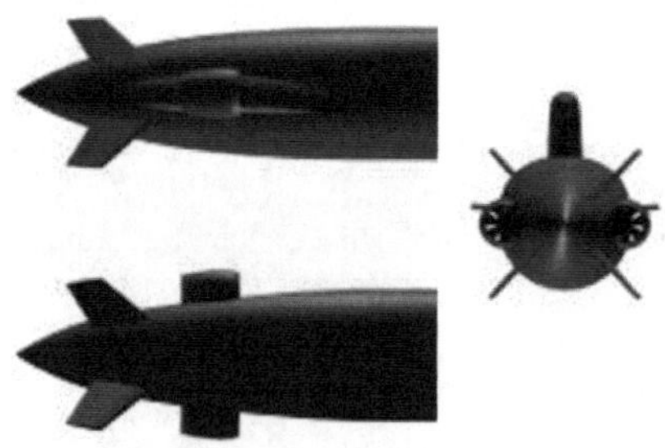

Figure 1: Design option 1

Figure 1 shows an initial concept design, where the off-axis propulsion, conceptually Rim drives, provides for a fully integrated axially located aft payload capability, The use of Rim drives gives freedom on their location and integration in the final design, facilitating investigation of a novel technology and providing a wider range of design options to consider, see [12]. The X-plane design looks to minimise the effect of the off-axis propulsor units on the aft control surfaces. The disadvantage of this design is that the external propulsion pods might be vulnerable to a shock event.

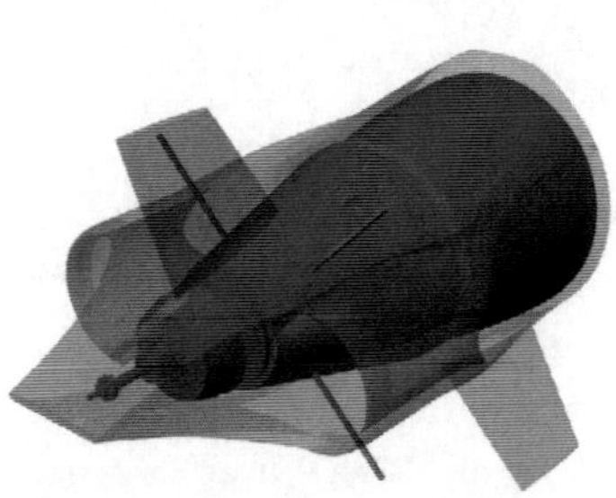

Figure 2: Design option 2

Figure 2 shows a design iteration again with an integrated aft payload bay configuration with outboard axial Rim drive units but with improved shock protection of the propulsion units. The part of the figure in red shows potentially how the design might integrate with the QinetiQ free-running model capability. This shows that there is not much space in the aft part to be able to wrap around the free-running model capability.

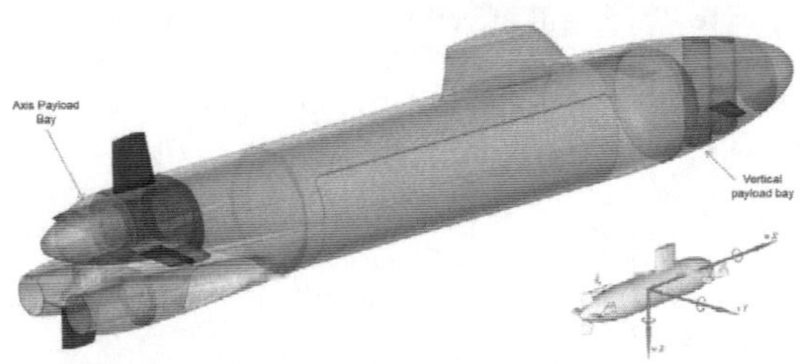


Figure 3: Design option 3 and axis system

The next iteration was Design option 3, which is shown in Figure 3. This has a payload capability both forward and aft and off-axis Rim drive units more integrated as part of the hull form design. For convenience, the duct size for the Rim drive is 3.5m in diameter.

This design has additional benefits, from an experimental perspective, by providing greater free volume aft, in which to house servos and other equipment used for model tests. The overall length of this design is 86m and it was this design that was taken forward to be evaluated by QinetiQ.

Early analysis, performed by BAE, of this concept included:

- Weight and buoyancy
- Resistance and powering predictions
- Stability and control

## 3. PRELIMINARY DESIGN EVALUATION

A series of preliminary design evaluation studies were conducted using Computational Fluid Dynamics (CFD) techniques prior to engaging in a model build programme. The purpose of these studies was to:

- Calculate the required propulsion thrust to size the propulsion system for the captive model experiments.
- Derive the stability indices.
- Assess the forces and moments on the concept design.

Since CFD calculations were to be performed at model scale Reynolds, the first decision was to choose the scale of the model to be built. The choice was based on the availability of Commercial-Off-The-Shelf (COTS) thrusters that could be used to emulate the Rim drives. The thruster units needed to be able to provide the required thrust but had to be integrated into the propulsion duct geometry; the size of the model also had to be appropriate for the planned hydrodynamic tests.

CFD flow calculations were performed using STAR-CCM+ numerical flow solver supplied by Siemens. The calculations used a three-dimensional domain surrounding the model that extended approx. 1.5L ahead of the forward perpendicular to 1.9L downstream of the aft perpendicular, 0.9L to port and starboard of the axis and 0.9L above and below the axis, (L is the model length). Mesh refinement was added to model the wakes. The effects of turbulence were modelled using the SST (Menter) k-Omega Turbulence model. The flow regime was assumed to be fully turbulent throughout the domain. Simulations were run in the unpropelled and the self-propulsion condition.

Flow calculations were performed for a range of state variables including pitch, drift, pitch rate, yaw rate and plane angles. All calculations were performed at a model scale speed of 2m/s; with forces and moments non-dimensionalised as follows:

$$ND\ force = \frac{force}{1/_2 \rho U^2 L^2} \qquad ND\ moment = \frac{moment}{1/_2 \rho U^2 L^3}$$

Initial calculations were performed with the model straight and level to understand the resistance of the body and the heave force and pitch moment on the body at zero angle of incidence.

| | **Non dimensional forces and moments (x 1000)** | | |
|---|---|---|---|
| **Condition** | **X'** | **Z'** | **M'** |
| Unpropelled | -2.22 | -0.78 | -0.33 |
| Self-Propulsion | 2.29 | -0.90 | -0.43 |

Table 1: Forces on body in straight and level flight.

The self-propulsion condition was modelled using momentum transfer through the actuator disks to obtain the required thrust at 2m/s; the calculation was performed iteratively with initial estimates being the thrust required to balance the predicted drag force.

Of interest is the difference in pitch moment between unpropelled and self-propulsion condition; as is to be expected the off-axis propulsion system is creating a positive pitch moment; the negative Z force is probably due to the presence of the bridge fin and interaction effects between the stern planes and the propulsor ducts.

The CFD computations were used to obtain the stability indices, [3] of the design. The derivation of these indices requires the so called linear derivatives to be calculated from the slopes of the pertinent forces at the origin. Having obtained these linear derivatives, the vertical and horizontal stability indices Gv and Gh can be calculated as follows (also shown are the values for this design).

$$G_v = \left(1 - \frac{M_w'(m' + Z_q')}{M_q' Z_w'}\right) \quad \text{and} \quad G_h = \left(\frac{1 + N_v'(m' + Y_r')}{N_r' Y_v'}\right)$$

In the case of this design both Gv and Gh were positive which is the accepted criteria that indicate vertical and horizontal stability respectively.

An interesting feature is the forces on the body due to the rudder being at an angle. Figure 4 shows the forces on the body (for the unpropelled case) when the rudder is applied, there is a significant negative heave force due to the rudder that will in turn create a bow down pitch moment. This heave force is actually larger than the sway force induced by a rudder angle; this rather undesirable effect warrants further examination.

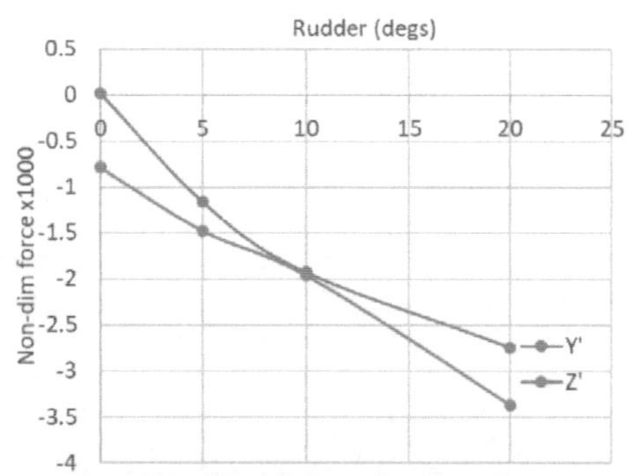


Fi

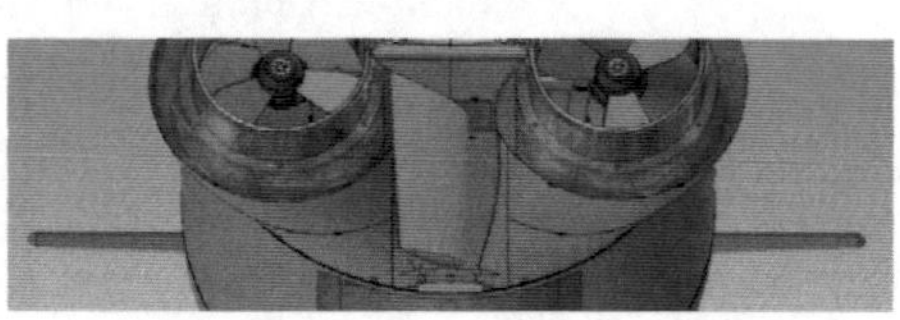

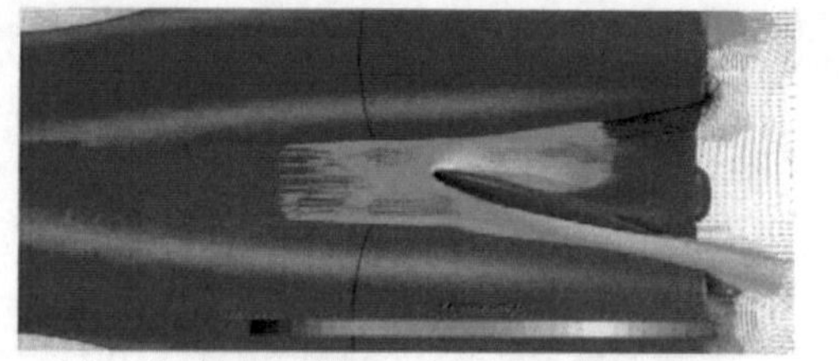

Figure 5: Blockage effects due to rudder angle at 20 degrees.

There is reduced flow over the lower rudder, created by the rudder blocking the flow between the underside of the two thruster ducts, Figure 5. This reduced flow will reduce the effectiveness of the rudder in the horizontal plane but is also the cause of the undesirable heave force due to the increase in the local pressure.

This design issue was considered sufficiently significant to warrant a modification prior to embarking on a physical model build programme. The design modification introduced a non-moving skeg as part of the lower rudder as shown in Figure 6. The outreach of the moving portion of the lower rudder was increased a little to ensure it was still effective; this meant that the lower rudder did extend below the keel significantly, which is unusual as it makes the lower rudder vulnerable in the event of grounding or operations in shallow water.

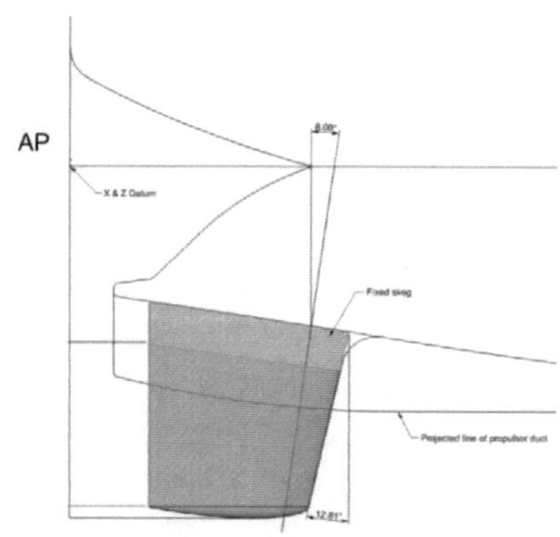


Figure 6: Modified lower rudder

## 3. CAPTIVE MODEL

### 3.1 MODEL BUILD

A model of the concept design was manufactured from Glass Reinforced Plastic (GRP). The internal structure of the model used a core cylindrical centre section that was fixed; the casing, bridge fin, bow and stern sections were then attached at different orientations to perform tests in either the vertical or the horizontal planes. The RimIM drives where in effect a set of COTS brushless thrusters from Technadyne. Clearly, the model had 2 units, one running a right hand and the other a left hand propeller.

Each of the units was capable of developing 220N of thrust, which based on the CFD predictions appeared to be sufficient to test the model at typical captive model test speeds.

The most complex aspect of a captive model is the means of actuating the stern control surfaces and measure the feedback demands. In this case a series of worm and wheel servo drives were used with independent feedback provided by potentiometers.

As mentioned previously this task was set to challenge the submarine hydrodynamics capability in QinetiQ, which includes model build. Several new techniques for captive model manufacture were established including 3D printing; as a result of the complexity of design, especially around the propulsor ducts, see Figure 7(a) for example. The traditional approach of GRP fabrication would have required several components to be made and assembled to produce the complex shape that sat between the aft part of the two ducts. Instead, rapid prototyping techniques were used to 3D print the complex shape as two components with a printed join for easy assembly.

(a)Thrusters (b) servo arrangement

(c) Complete model

Figure 7: Design detail of captive model

Figure 7(b) shows the complexities of the actuation mechanism (and feedback) used in the captive models test. Figure 7(c) shows the completed model and Figure 8 shows the model mounted inverted on the Rotating Arm at QinetiQ

## 3.2 MODEL TESTS

The concept of a constrained model experiment is simple and well-established [4]. Internal instrumentation consists of a strain-gauge balance capable of measuring the forces and moments acting on the body in all six degrees of freedom. The experimental design is fundamentally a function of the form of the coefficients; that is, the experiment must be designed so that the state variables can be controlled as independent variables. The model is towed in a towing tank at a range of body incidences in both the horizontal and vertical plane; thus providing the relationship between the forces, moments and the velocities. Then, with the model at zero angle of incidence, the various control surfaces are moved over their working range to measure forces and moments generated as a result of their deflection. The experiment then includes oscillation tests where the model is subjected to sinusoidal motion. In these tests, it is not a single parameter that is varying as the oscillation will induce components due to both velocity and acceleration.

Figure 8: Model on Rotating Arm

Therefore, to determine the relationship between forces and moments and accelerations requires the components due to velocities to be removed (using the terms derived from the steady state tests). Once the towing tank experiment is complete, the model would then be transferred to a rotating arm where the forces and moments due to rotational velocities are established, along with further tests that include the various control surfaces at angles of attack.

These experiments generate a database of forces and moments acting on the submarine, as functions of the particular attitude to the flow and the deflections of the control surfaces. Once these data have been captured they are combined and a full least-squares regression analysis is performed on the entire data to provide the derivatives that can then be used directly in the equations of motion to reproduce the forces and moments [5]. In the first instance the standard model in [6], was considered, the form of the force equations were then modified to reflect features of this particular design.

## 3.3 DATA ANALYSIS

Analysis of the experimental data confirmed that, in most cases, the standard hydrodynamic coefficients in [6], were a good fit to the experiment data, although a number of additional terms were introduced to improve the fit. The following equation shows the contribution to heave force from the independent state variables.

$$
\begin{aligned}
Z' = & \left(Z'_{uu}u'^2 + Z'_{uw}u'w' + Z'_{uv}u'v' + Z'_{vv}v'^2 + Z'_{uu\delta b}u'^2\delta b + Z'_{uu\delta s}u'^2\delta s\right) \\
& + \left(Z'_{wv}w'v' + Z'_{u|w|}u'|w'| + Z'_{|wv|}|w'v'|\right) + \left(Z'_{\dot{w}}\dot{w}' + Z'_{uq}u'q' + Z'_{vp}v'p' + Z'_{vr}v'r'\right) \\
& + \left(Z'_{wv\left|\frac{q}{w}\right|}w'v'\left|\frac{q'}{w'}\right| + Z'_{u|q|\delta s}u'|q'|\delta s\right) + \left(Z'_{\dot{q}}\dot{q}' + Z'_{pp}p'^2 + Z'_{rr}r'^2 + Z'_{rp}r'p'\right) \\
& + (m'g' - B')\cos\phi\cos\theta + \left(Z'_{uw\eta}u'w' + Z'_{uq\eta}u'q' + Z'_{w|w|\eta}w'|w'| + +Z'_{uu\delta s\eta}u'^2\delta s\right)(\eta - 1)
\end{aligned}
$$

The reader should refer to [7] for an explanation of the formulation of these equations. Fitting a curve of this form to the measured vertical force against pitch rate shows an example of where introducing additional coefficients did improved the fit.

During the analysis, there was evidence that the standard coefficients used to fit to measured $Z'$ as a function of pitch rate (linear terms $Z_{UQ}'$ and $Z_{UQn}'$) would benefit from the addition of a second order term, $Z_{QQ}'$.

Also shown in Figure 8 is the effect on the vertical force due to the propulsor state or n' is defined as the following:

$$n' = \frac{Instantaneous\ rpm}{rpm\ for\ self\ propulsion\ at\ instantaneous\ speed}$$

So, whilst the submarine is under braking, applying astern RPM to slow down, n' will be negative; the approach used by [6] to model the effects of the propulsor state is by introducing further coefficients that are linearly dependent upon n'. Hence, a captive model test plan would ordinarily include tests that involve varying the value of n'.

When analysing the tests where n' = -1, it became apparent that, the variation in the measured forces and moments due to varying n' was not consistent with that expected according to the standard mathematical model. In Figure 9, the grey points deviate significantly from the data points measured for the other values of n'.

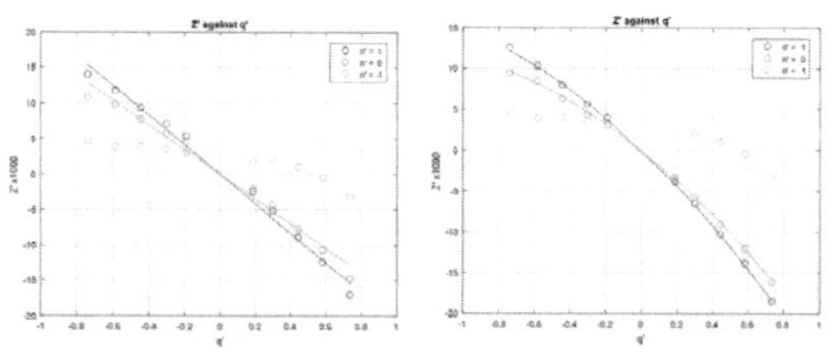

(a) Linear terms only (b) Second order terms

Figure 9: Fit to heave force using linear and second order terms in q'

Establishing a revised mathematical mode to account for the change in forces on the body due to varying the propulsion state would likely require data more tests to be performed at a greater number of propulsion states.

In the absence of these additional data, to ensure that the quality of the fit across the other propulsion states (n' = 1 and 0), the data for the case of n' = -1 were disregarded from further analysis. This simplification means that the extracted coefficient set is unlikely to predict braking manoeuvres with any degree of accurately; we shall see from the free-running model tests the implications of this limitation.

It was found from the analysis of the captive model tests that it was unnecessary to develop a new propulsion model to represent the thrust from the Rim drives. Instead, it was sufficient to adopt a method similar to that in [6]. The propulsion model in [6] is based on 9 coefficients, $B_1$ to $B_9$, where three sets of three coefficients were used to define a quadratic through segments of the propulsion curve depending upon n'.

The modifications required for the off-axis propulsion system was to deal with the case where one or two thrusters are used. So, an additional variable thrust state was introduced.

$$thrust\ state = \frac{\frac{1}{2}\,(port\ rpm + starboard\ rpm)}{\max(port\ rpm, starboard\ rpm)}$$

(=1 for both thrusters operating and 0.5 for one only operating).

For the case, where a single thruster is operating an additional yaw moment is generated given by:

$$N_{port} = \frac{port\ rpm}{port\ rpm + starboard\ rpm} \times X_n \times lever\ arm$$

$$N_{stbd} = \frac{stbd\ rpm}{port\ rpm + starboard\ rpm} \times X_n \times (-lever\ arm)$$

Where the lever arm is the lateral distance between the centre of the thrusters and the axis line and $X_n$ is the total thrust determined from the propulsion model. The pitch moment, due to the off-axis nature of the thrusters (vertical plane) was accommodated by the existing $M_{UU}'$ term in the pitch moment equation from [6].

## 4. FREE-RUNNING MODEL

### 4.1 MODEL DESIGN

A free-running model (FRM) was constructed at the same scale as the captive model. This enabled the re-use of the entire thruster section from the captive model, shown in Figure 10 (the starboard thruster tunnel is hidden to show the location of the thruster itself).

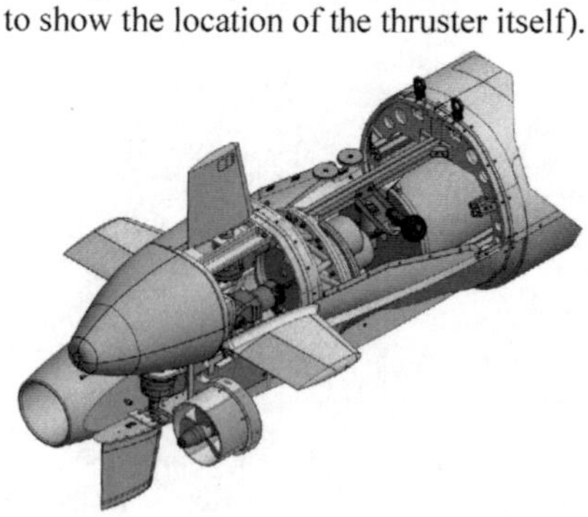

Figure 10: Aft section of the free-running model

The solution to the free-running model was to use the Submarine Research Model II (SRMII) concept, described in [8]. The system for actuating the control surfaces is different in the captive model compared to the FRM, moving from a worm and wheel arranged to Heasons servo units respectively. The SRMII was fitted with its standard instrumentation package, including a Fibre Optic Gyro (FOG), a Doppler Velocity Log (DVL) and forward and aft pressure depth transducers; also included were the moveable mass and ballast systems described in [9]. Each thruster was driven by its own independent drive controller, receiving digital command from the model control system.

The depth and course autopilots used in the free-running model tests were not representative of any particular control algorithms, but were controllers that are routinely used by QinetiQ for their model test programmes. The depth controller was a multivariable type based on an H-infinity ($H\infty$) design process; the course controller algorithm was a generic phase advance controller, with a gain and time constant set for a particular speed.

The flood recovery tests used a specific depth controller that looked to emulate a typical operator response to such a casualty, which assigned the bow planes to respond only to depth error and the stern planes to control the pitch angle during the recovery following a simulated flooding.

## 4.2 RUN PROGRAMME

The run programme consisted of the following:

### 4.2.1 Set-Up Runs

Preliminary runs were conducted to confirm the thruster rpm to speed profiles, calculate the balance angles, and conduct roll incline tests to confirm the BG of the model.

### 4.2.2 Turning Circles

Turning circles were performed to measure parameters such as Advance, Transfer and Tactical Diameter. Turns were conducted to both port and starboard, and with the use of twin and single thrust.

### 4.2.3 Free Turns

These manoeuvres are open loop turning circles, i.e., with no depth control where the bow and stern planes were held fixed at the approach angles. Depth and pitch limits were in place, beyond which the hydroplanes would be released and depth control regained.

### 4.2.4 Horizontal Zig-Zag Manoeuvres

These standard manoeuvres are conducted to demonstrate the yaw-rate checking ability, as well as for simulation validation.

### 4.2.5 Vertical Zig-Zag Manoeuvres And Pulses

These were performed in three modes: stern planes acting on a pitch trigger; stern planes acting on a depth trigger; and bow planes acting on a depth trigger. Again, these runs are used to characterise the dynamic of the submarine in the vertical plane for use in simulation validation.

### 4.2.6 Open And Closed Loop Depth Changes

Depth change manoeuvres were conducted to explore the effect of the bow plane deflection on the angle of attack during a constant pitch angle rise/dive.

### 4.2.7 Hydroplane Jam Response Manoeuvres

These runs applied a fixed jam angle to one of the two stern planes, while the remaining control surfaces continued under autopilot control.

### 4.2.8 Flood Recovery

From a slow speed initial condition, a flood was simulated, the response was to order a depth change to the surface and to accelerate to a target speed.

### 4.2.9 Differential Thrust

The opportunity was taken to apply ahead thrust to one thruster and astern thrust to the other, thus generating a yaw moment while effectively hovering.

## 4.3 FURTHER DEVELOPMENT OF THE MATHEMATICAL MODEL

As discussed earlier, the CFD studies had identified that the original lower rudder arrangement could lead to a significant bow down pitch moment during a turn. This led to the design change shown in Figure 6. However, the free-running model did still exhibit a pitch response on entry to a turn, it was therefore necessary to improve the mathematical model to be able to simulate this effect; two additional coefficient were included, $Z_{UU|\delta R|}'$ and $M_{UU|\delta R|}'$. Furthermore, with these additional terms in place the existing cross-coupling terms required revision, and this led to the introduction of the non-linear terms $Z_{U|V|}'$ and $M_{U|V|}'$.

## 4.4 SAMPLE RESULTS

Clearly, there are a significant amount data that could be presented, the following represents a only a small selection. Turning circle manoeuvres were conducted in both the Haslar Ocean Basin and at a deep water reservoir.

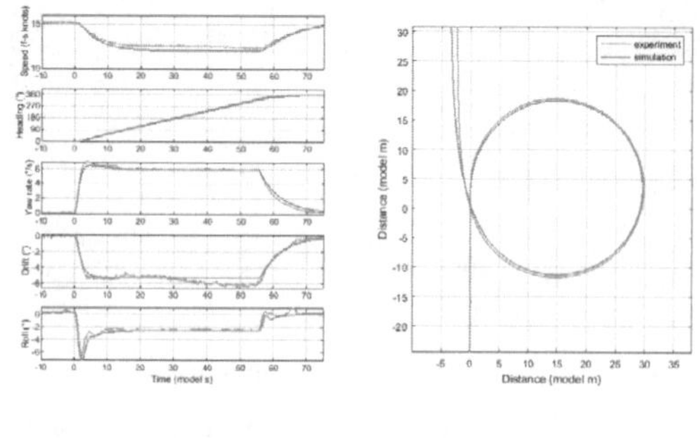


Figure 11: Turning circle and pull-out (Experiment and simulation)

It is clear from Figure 11, that this particular design was very stable in the horizontal plane; as seen by a turning circle around 6 times the submarine length. Also, following the pull-out part of the manoeuvre the submarine quickly adopts a constant heading meaning the yaw rate reduces to zero shortly after setting the rudder to midships. Of note, is that this design rolls away from the turn (both snap roll and steady roll) unlike more traditional design, chiefly due to the upper rudder being more effective than the lower; this might be a contributing factor in the large turning circles.

Recall that one of the issues with this design was the pitch moment during a turn. In fact, pitch and depth control during a turn were not an issue, with excursions being insignificant throughout the manoeuvre

Single hydroplane jams runs were performed by letting the model reach a brief straight and level flight steady state condition, then "jamming" one of the aft control surfaces to a fixed angle. As part of the recovery process, the remaining control surfaces where under autopilot control to maintain depth, in these cases the RPM on both thrusters was kept constant. Hydroplane jams were conducted for a range of jams to rise and dive (both port and starboard) at a number of initial speeds with the twin thrusters or a single thruster.

In all cases, the pitch and depth excursions were negligible and this was reflected in the simulations of these manoeuvres. Figure 12 plots two examples of stern plane jams to rise, showing good simulation agreement with the depth and pitch responses, as well as the control surface angles adopted.

However, it should be noted that the trajectories in Figure 12 are for only small depth (less than 0.05m model-scale) and pitch ($\approx 1.1°$) excursions. The trajectories discussed in Section 5.1 result in significantly larger excursions.

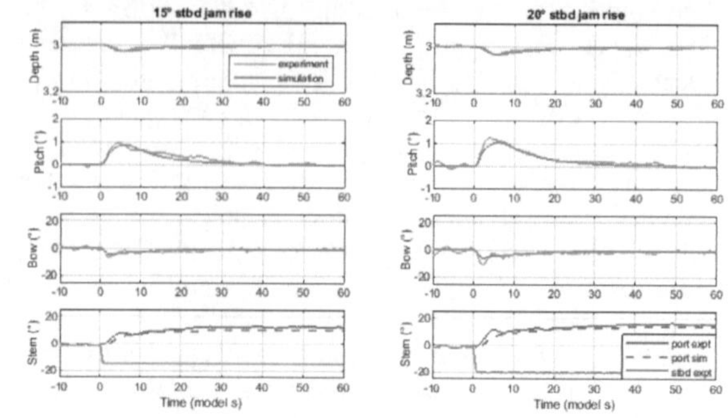


Figure 12: Jams to rise (Experiment and simulation)

A series of runs were conducted to simulate a flood in the aft of the submarine (using the ballast can capability in SRMII). These tests represented a flood mass of around 20-30 tonnes. All runs had a low initial forward speed, the response to the flood including applying thruster recovery actions shortly after initiating the flood and using the control surfaces to maintain pitch during the ascent. Figure 13 shows, at full-scale, the response of the submarine to a 30 tonne flood at 3 knots initial speed. The major difference between the simulation and experiments is that it appears that pitch is higher in simulation, even though the control surface demands are the same as the experiment data; the physical model did not appear to have any issues controlling pitch.

In the recovery process, the bow planes set to full rise in both the experiments and simulation. However, since stall effects are not modelled in simulation, the bow planes are too effective and cause this larger pitch angle during recovery. In the experiment, it is likely that the bow planes are experiencing stall, and therefore not generating the same pitch moment.

Some runs were conducted which brought the model into a near-stationary hovering mode, and then differential propulsion was applied with ahead rpm to one thruster and astern to the other. The effect of this was to generate a yaw moment that was sufficient to turn the model turning around. The yaw rate was controlled, i.e. stopped and reversed, by switching the two thruster directions. It was demonstrated that the model could be turned around almost on the spot. Figure 14 plots snapshots of the submarine's location at 2 minute intervals (all full-scale).

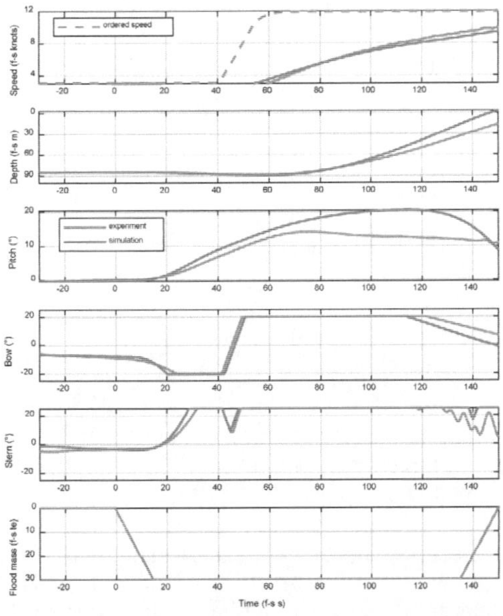


Figure 13: Flood recovery (Experiment and simulation)

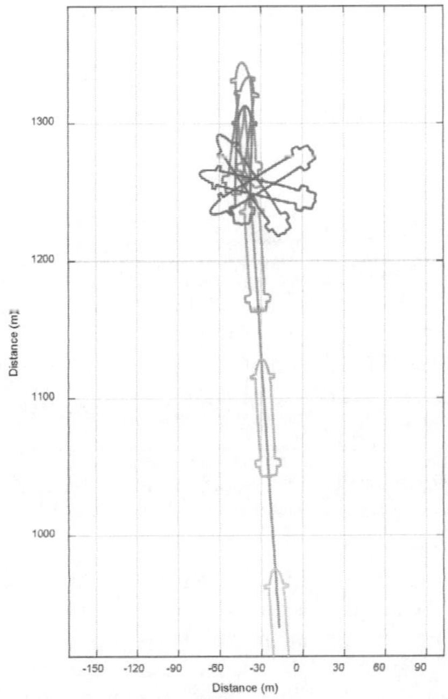


Figure 14: Differential thrust

In Figure 14, the colour of each submarine outline intensifies over time. The submarine approaches from the bottom of the figure at a speed of 3 knots, coming to a stop towards the top of the figure. There is a small deviation in heading to starboard, before the differential thrust is set and the submarine rotates to port. Although only the first 120° change of heading is shown, the manoeuvre continued for a full 360° turn.

## 5. PRELIMINARY SAFE OPERATING ENVELOPES

The final step in this process is to understand the form that any provisional operator guidance, to mitigate the risk to the submarine following a control surface jam or flooding incidence casualty scenarios, would look like. A Safe Operating Envelope (SOE) diagram, see Figure 15, provides details of safe combinations of speeds and depths at which the submarine can operate, should an incident occur. The format of the SOE for this off-axis hull form will be driven, to some extent, by the design itself but also by the recovery methodologies that are used as a consequence of having a twin propulsion system. In the case of this novel off-axis design, the system arrangements may mean that recovery methods are quite different to a typical on-axis propulsion system.

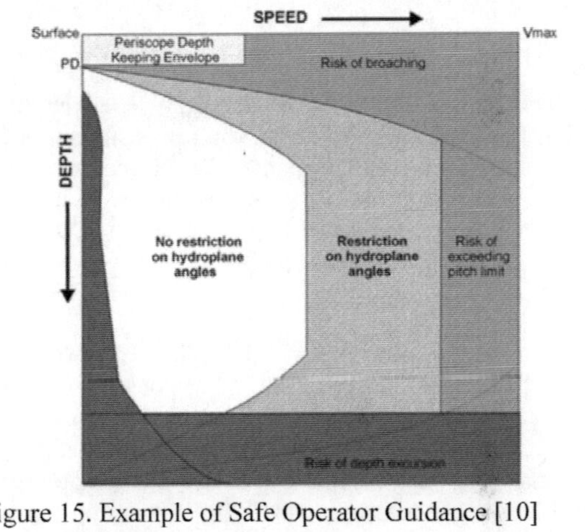


Figure 15. Example of Safe Operator Guidance [10]

There were a number of assumptions made to support the development of the SOE;

- Maximum speed is set at 16 knots.
- The Maximum Excursion Depth is 224m, based upon the average collapse depth of World War II submarine designs, [11].
- A single stern plane is considered to be a credible failure.
- Jam recoveries will use the existing working control surfaces using a manual recovery process.
- Twin propulsion will be available throughout the recovery.

### 5.1 PLANE JAMS

The most appropriate recovery action for a control surface jam is to reduce speed, either by applying astern RPM (braking) or by reducing RPM (decelerating). In the case of this concept design, the captive model data suggested that under astern RPM, a significant pitch moment would develop; this was confirmed to be the case in the free-running model tests. For a jam to dive, applying astern RPM caused a significant positive pitch response but the model still experienced an increase in depth and did not recovery safely from the jam. As astern RPM, the thruster will be attempting to push fluid back up the thruster tunnel, creating complex unsteady flow around the inlet to the thruster tunnels. This complex flow creates these unexpected forces and moments on the body. Therefore, deceleration, rather than braking, was considered to be the most suitable response to a jam.

The ability of the submarine to recover from a stern plane jam tends to be sensitive to the speed that is desired during the recovery phase; for this design, 4 knots was considered the best recovery speed, based on maximising the operational area. Another parameter that the operating envelope is sensitive to are the recovery angles that are demanded of the working stern plane and the bow planes. In this case, a bow plane angle of 20° and the working stern plane set to 5° in the opposite direction to the jammed stern plane seemed to provide the best outcome.

This would appear to be counter-intuitive, given that setting the working stern plane to an opposite angle to the jammed stern plane appears to be the most obvious choice of recovery. However, in this design, it appears that the bow planes have a greater level of control authority (compared to the stern planes) than is typical of submarines.

Based upon the outcome of the preliminary work to establish the most suitable form of response to a plane jams, an SOE was calculated in the pitch limited Safe Manoeuvring Envelope (SME) format; this used a range of initial pitch angles and a constant 25° single stern plane jam angle. The resultant SME is shown in Figure 16.

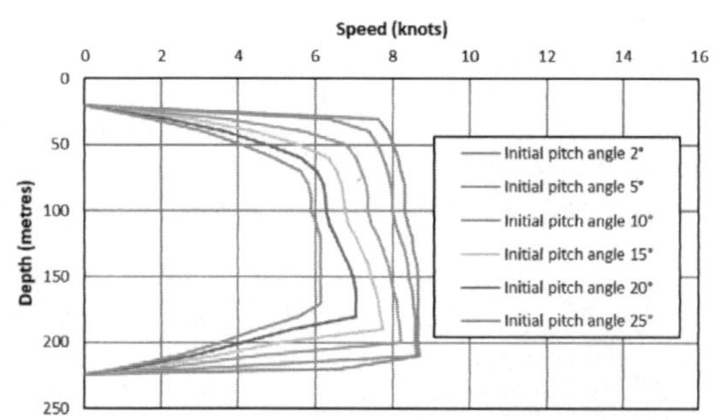


Figure 16. Pitch limited SME at 25° jam (single stern plane) for the concept design.

Figure 16 shows that this concept design submarine is severely limited in its operational area compared with a traditional submarine. For comparison, a coefficient dataset representative of a more traditional submarine design was used to calculate an equivalent SOE This used the same twin thruster propulsion model, but with different manoeuvring coefficients. The SME, shown in Figure 17, for the traditional submarine is able to operate up to its maximum speed of 16 knots with up to a 10° initial pitch angle following a 25° stern plane jam, whereas the concept design cannot operate faster than 9 knots for an equivalent condition.

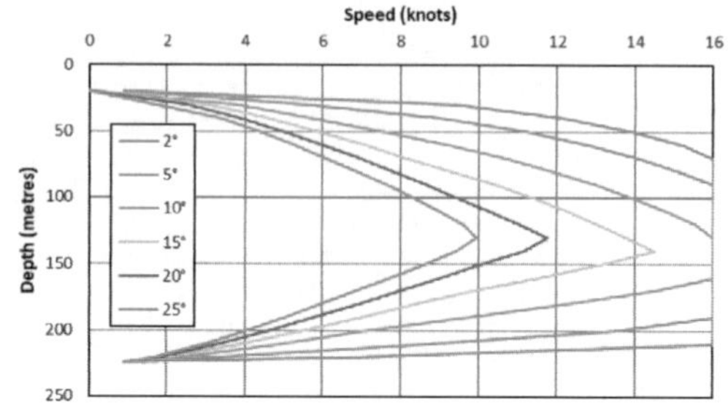


Figure 17. Equivalent pitch limited SME (single stern plane) for a traditional submarine design.

Therefore, to expand the potential operational area of the SOE of the concept design, the format of the SOE was changes from a pitch limited Safe Manoeuvring Envelope (SME) to a plane limited Manoeuvring Limitation Diagram (MLD). In this format the initial pitch angle remains constant (2° for this example) and the responses to a range of jam angles are calculated. A depth keeping MLD for the concept design is shown in Figure 18. It should be noted that the 2° initial pitch angle line in Figure 16 is the same result as the 25° jam angle line in Figure 18.

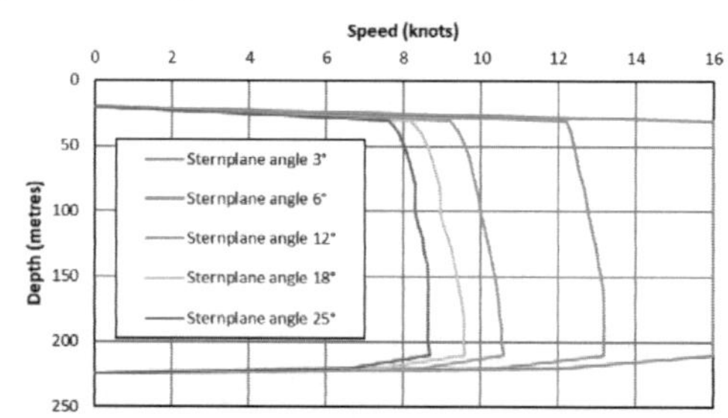


Figure 18. Plane limited MLD at 2° initial pitch (single stern plane) for the concept design.

As can be seen in Figure 18, when a 3° jam angle is considered, the available operational area for the concept submarine includes its maximum speed. However, this has consequences as to operate at 16 knots, the planesman or autopilot would need to remain within a ±3° limit on the stern plane movement.

So, why is the operational envelope of the concept design so limiting? Closer inspection of the jam to rise trajectories shows that failures at the lower speeds are due to exceeding the minimum depth criterion (i.e., the submarine is broaching. For the jams to dive, even when there is an initial bow down pitch angle, the trajectories are failing on the bow up pitch limit.

 

In the case where the submarine starts at a shallow initial depth, a significant depth excursion to dive would be required to before the submarine fails by reaching the maximum excursion depth. In the case of the higher speeds, a significant positive pitch angle develops as a consequence of the vertical velocity on the body following the stern plane jam to dive. Therefore, the pass/fail boundary for jams to dive are due to the trajectory failing on a bow up pitch angle before the rate of change of depth can be arrested.

As mentioned previously, since the control authority of the bow planes, compared to the stern planes, is larger than more traditional designs, it was felt prudent to calculate and SOE for bow plane jams also. A pitch limited SME for a linked bow plane jam (i.e. port and starboard planes together) of 20° was determined assuming stern plane recovery angles of 25°. However, whilst the jam to rise SME results showed that the concept design could be operated up to 16 knots with an initial pitch of 2°, the jam to dive trajectories all failed on exceeding maximum excursion depth. This indicates that there is some asymmetry within the vertical plane response of this concept design.

### 5.2 FLOODS

In the case of a flood, the first thing to identify was what represented a critical flood case for this concept design. A nominal 0.4m diameter flood hole penetration located in the Main Machinery Space was considered as the limiting flood scenario. As a consequence of this type of flood the submarine would experience power loss 10 seconds after the hull valve was shut to secure the flood; an alarm would be activated as the flood mass exceeds 10 tonnes. The standard response to a flood alarm would be to blow main ballast tanks, increase the speed of the submarine and use the planes to control the ascent. The calculated Flood Avoid Zone for the concept design is shown in Figure 19 compared with an equivalent FAZ for the traditional submarine design.

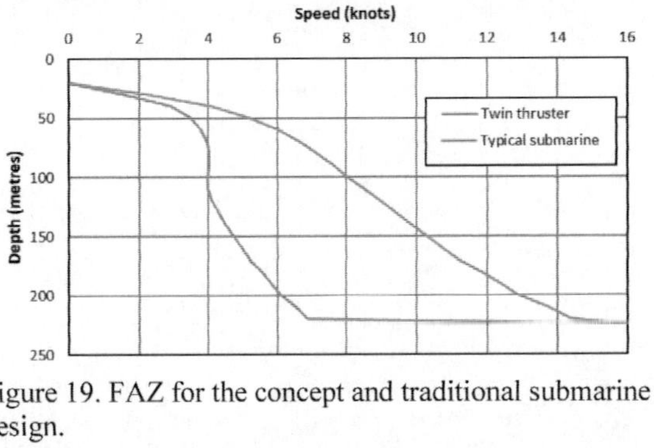


Figure 19. FAZ for the concept and traditional submarine design.

This shows that the concept design is more capable of recovering from an equivalent flood at slower speeds that the traditional design for the same plane angles, propulsion and blowing model. This is largely due to the control authority available from the bow planes; in the case of the concept design, the bow planes are capable of generating a pitch moment that can counter the flood mass more effectively than the traditional design. As such, the pitch angle can be controlled more effectively during the recovery from the flooding incident.

## 6. CONCLUSIONS

This paper has described a challenging design and hydrodynamic assessment exercise; in which traditional assessment techniques have been developed to suit a novel submarine design with off-axis propulsion.

The focus of the design was to provide a hull form with an enhanced launch and recovery capability of various payload configurations and a novel off-axis propulsion system was identified as a technology enabling strategy that could facilitate an enhanced payload design and integration solution.

Moreover, this design has gone through a rigorous assessment process to understand the impact of these novel features on hydrodynamic performance. This design has been evaluated using of CFD techniques that were able to identify design issues, i.e. the all moving lower rudder situated between the two thruster tunnels, early in the model design process that prevented this unsuitable solution from going forward to model manufacture.

A series of captive model tests have been completed to generate a database of forces and moments acting on the submarine, as functions of the particular attitude to the flow and the deflections of the control surfaces. From these data, an extension to the typical quasi-steady state coefficient based mathematical model has been proposed.

This mathematical model has been validated using data generated from free-running model tests; the QinetiQ reconfigurable free-running model capability, SRMII, was configured to the concept design. A wide range of tests were conducted ranging from simple propulsion tests to modelling flood recoveries.

Finally, using this validated mathematical model, the paper proposes a preliminary SOE. It appears that the control authority of the bow planes compared with the stern planes is a critical design feature. Over sized bow planes may help with flood recoveries but can have a negative impact on recoveries from plane jams – particularly when relying on manual control during these scenarios.

This work has provided an opportunity to de-risk novel design features to inform future underwater platform programmes and by undertaking a full manoeuvring and control evaluation of this design, the traditional testing techniques have been challenged and developed.

## 7. ACKNOWLEDGEMENTS

The authors would like to acknowledge that the work reported here represents a summary of a significant amount of work conducted by the Hydrodynamics Team in QinetiQ Haslar.

The authors also thank the UK MoD for supporting this work.

## 8. REFERENCES

1. CROSSLAND P, BAYLISS J, KIMBER N, 'Submarine hull form design, safety and operational guidance, *Submarine Institute Australia Technology Conference,* 2013, Adelaide.

2. CROSSLAND P, MARCHANT P, THOMPSON N, "Evaluating the Manoeuvring Performance of an X-Plane Submarine", *Warship 2011: Naval Submarines and UUVs, Bath,* June, 2011.

3. BISHOP, RED, PARKINSON, AG, "Directional stability and control of marine vehicles", *Department of Mechanical Engineering, University College London, Nav. Arch. Rep. 3/70*, 1970.

4. LEWIS, EV, "Principles of Naval Architecture", *Vol III, Motions in waves and controllability, SNAME*, 1989.

5. KIMBER, NI, CROSSLAND, P, "Advances in experimental techniques for understanding the manoeuvring performance of submarines", *RINA Warship 2008: Naval Submarines 9, Glasgow,* June 2008.

6. GERTLER, M, HAGEN GR, "Standard equations of motion for submarine simulation", *Naval Ship Research and Development Center Report No. 2510, David Taylor Model Basin, Carderock, MD, USA,* June 1967.

7. RENILSON M, "Submarine Hydrodynamics", *Springer International Publishing, ISBN 978-3-319-79056-5,* 2018

8. P CROSSLAND, R C NOKES, S DUNNINGHAM, P MARCHANT, N KIMBER, "SRMII - A Reconfigurable Free Running Model Capability For Submarines With Large L/D Ratios", *RINA Warship 2014: Naval Submarines & UUV's, Bath, UK*, 18-19 June 2014.

9. CROSSLAND P, MARCHANT P, "Simulating flood recovery manoeuvres using a free running submarine model", *PACIFIC 2019, International Maritime Conference, Sydney,* October 2019.

10. BAYLISS, JA, KIMBER, NI, MARCHANT, P, BRYARS J, *Submarine trials and experimentation – dealing with real-life data,* RINA Warship 2005: Naval Submarines 8, London, 2005.

11. GABLER U., *Submarine design*, Bernard & Graefe Verlag, Koblenz, ISBN 3-7637-0124-9, 1986.

12. COLLINS, L, WROBEL P, *The Current State of Rim Drive Technology and its Application to Submarine Technology*, RINA Warship 2014: Naval Submarines and UUVs, Bath 2014.

## 9. AUTHORS BIOGRAPHY

**Paul Crossland** currently leads the Submarine and Propulsion Team at QinetiQ Haslar. He is responsible for the delivery of the submarine hydrodynamics programme for the UK MoD, developing experimental and computational techniques for assessing the performance of submarines, including understanding the behaviour of submarine close to the free surface.

**Richard Crocker** is currently the Chief Naval Architect for a major submarine programme at BAE Systems – Maritime Submarine in Barrow in Furness

**Stuart Machin** is the current Head of Submarine Naval Architecture within the Naval Authority Group of the UK MOD Submarine Delivery Agency, where he has responsibility for naval architecture standards and design guidance and issuing certification for UK submarines.

# DEVELOPMENT OF A SUBMARINE MANEUVERING SIMULATOR

**E G Nascimento**, **E A Tannuri,** University of São Paulo, Brazil


## SUMMARY

This work consisted of the implementation of a numerical model of a Submarine in the Numerical Offshore Tank of the University of São Paulo (TPN-USP) based on the concepts of maneuverability of vessels, thus creating the first Submarine Simulator with fully Brazilian technology.


## NOMENCLATURE

| | |
|---|---|
| $m$ | Mass of submarine, including water in free flooding spaces (submerged mass) (kg) |
| $W$, $B$ | Weight and buoyancy of the submarine, including water in free flooding spaces (N) |
| $I_x, I_y, I_z$ | Moments of inertia of submarine about x, y and z axes (kg m$^2$) |
| $x_G, y_G, z_G$ | The x, y and z coordinate of the centre of gravity in relation the origin reference's frame (m) |
| $x_B, y_B, z_B$ | The x, y and z coordinate of the centre of buoyancy in relation the origin reference's frame (m) |
| $g$ | gravity acceleration (m s$^{-2}$) |
| $\overline{BG}$ | form stability modulus (m) |
| $l$ | length between perpendiculars (m) |
| $\rho$ | water mass density (kg m$^{-3}$) |
| U | Velocity of origin of body axes relative to fluid (m s$^{-1}$) |
| $u$ | Component of U in the direction of the x-axis (ship's reference frame) (m s$^{-1}$) |
| $v$ | Component of $U$ in the direction of the y axis (ship's reference frame) (m s$^{-1}$) |
| $w$ | Component of $U$ in the direction of the z-axis (ship's reference frame) (m s$^{-1}$) |
| $\dot{u}$ | Time rate of change of $u$ in direction of the x-axis (m s$^{-2}$) |
| $\dot{v}$ | Time rate of change of $v$ in direction of the y-axis (m s$^{-2}$) |
| $\dot{w}$ | Time rate of change of $w$ in direction of the z-axis (m s$^{-2}$) |
| $u(x)$, $v(x)$, $w(x)$ | Velocity component of any x coordinate (e.g., $w(x) = w - xq$) (m s$^{-1}$) |
| $\theta$, $\psi$, $\phi$ | Angle of pitch, yaw and roll (rad) |
| $p$, $q$, $r$ | Angular velocity component about axes relative to fluid (roll, pitch, yaw) (rad s$^{-1}$) |
| $p, q, \dot{r}$ | Angular acceleration component about axes relative to fluid (roll, pitch, yaw) (rad s$^{-2}$) |
| $\delta_b, \delta_s, \delta_r$ | Deflection angle of control surfaces: bowplanes/sailplanes, sternplanes and rudder, respectively (rad) |
| $X, Y, Z$ | Hydrodynamic forces components along body's axes reference frame (N) |
| $K, M, N$ | Hydrodynamic moment components about body's x, y and z axes (N m) |
| $A(t)$ | Force or moment $A$ as function of time (A can be $X$, $Y$, $Z$ or $K$, $M$, $N$) |
| $A_b$ | Feldman's notation: Force or moment $A$ as function of $ub$, $b$ is a movement parameter of speed or acceleration (u, v, w, p, q, r, $\dot{u}$, $\dot{v}$, $\dot{w}$, $\dot{p}, \dot{q}$ or $\dot{r}$) or control surfaces angle ($\delta_b$, $\delta_s$ or $\delta_r$) |
| $A_b{}'$ | Feldman's notation: Nondimensional coefficient used in representing force or moment $A$ as function of $ub$ |
| $A_*{}'$ | Feldman's notation: Nondimensional coefficient used in representing force or moment $A$ as function of $u^2$ |
| $A_b$ | Spencer's notation: Force or moment $A$ as function of $b$, first order hydrodynamic coefficient |
| η | Feldman's notation: ratio of command speed and u. The command speed is the steady value of ahead speed component u for a given propeller rpm when body angle and control surfaces angles area zero |
| $C$ | Feldman's notation: Variable coefficient used in scaling model thrust and drag data to full-scale |
| $c_d$ | Feldman's notation: Coefficient used in integrating forces and moments along hull due to local cross-flow |
| $\bar{c}_L$ | Feldman's notation: Modified nondimensional sectional lift-curve slope used in computing the effects of the hull-bound vortex due to lift on the bridge fairwater |
| $b(x)$ | Feldman's notation: Local beam of hull in x-y plane |
| $\bar{v}_{FW}$ | Feldman's notation: Velocity component in the y-axis direction at the starting position of the hull-bound vortex due to lift on bridge fairwater |
| $\tau[x]$ | Feldman's notation: Time interval required for vortex to travel from any distance on x coordinate. |

## 1. INTRODUCTION

The TPN-USP Ship Maneuvering Simulation Center is accredited by the ITTC (International Towing Tank Conference). It is based on an in-house mathematical model and simulation software. It executes dozens of simulations annually that contribute to various engineering projects and operations related to ports and the offshore industry (Tannuri and Martins, 2018 [11]), the main ship bridge mock-up is shown in Figure 1. The numerical model of the TPN-USP Ship Maneuvering Simulator is based on vessels operating on the surface such as ships, tugs and offshore structures. The Center executes simulations of new vessels assessing ports, navigation viability, risk analysis and new port infrastructure. It also executes evaluation of offshore operations and vessels, such as oil rigs, DP tankers and PSVs.

The present work consisted in adapting the mathematical model of the TPN Ship Maneuvering Simulator, making the necessary changes to allow the simulation of submersibles and submarines. The final product is that the TPN Simulator, from 2020, is also capable of simulating submarines, allowing for various tests and analyses related to maneuverability and all the possibilities that this new tool offers.

Figure 1: Maneuvering Simulation Centre of TPN (source: TPN-USP)

## 2. THE MANEUVERABILITY

The maneuverability of a vessel is an essential aspect, especially in restricted and port waters for merchant ships and in operations for warships and submarines. In the case of warships and submarines, maneuverability has tactical and operational purposes of fundamental importance regarding the survivability of the vessel. Controllability is the study of this performance and is summarized in three main areas or functions:

- Coursekeeping, that is, the ship's capacity to maintain its heading. It is directly related to directional stability.
- Maneuvering, that is, the capacity for controlled change of heading. It seeks to assess how easy this change of heading is, as well as the radius and distance required to complete this change.
- Speed Changing, that is, the controlled change of speed, including stop or machines in reverse. Interest centers on the ease, rapidity and distance covered in accomplishing changes.

Therefore, there are several types of maneuvers, both to assess the vessel's performance with fixed controls or with active controls. For this reason, their execution and evaluation are of fundamental importance to understand the behavior and performance of the ship. Much about the theoretical basis of maneuverability can be found in the literature (Arentzen e Mandel (1961) [1], Feldman (1979) [2], Fossen (2011) [3], Renilson (2015) [6], Spencer (1968) [9]), however, a basic understanding of the subject is necessary to understand the aspects of each maneuver. The controllability parameters and stability indexes determine the behaviour of the ship or submarine in the various maneuvers. Thus, it is possible to assess the risks and possibilities of access to a port or harbor terminal, as well as in the case of more unfavorable or emergency environmental conditions or situations.

It is very advantageous to know in advance the vessel's behavior in each type of maneuver so that one can be aware of the capabilities and limitations of that ship in each situation and, thus, correctly decide which maneuver to perform or not.

In the case of submarines, the design and evaluation of the Safe Operating Envelope (SOE) is only well achievable using a proper submarine simulator. The SOE is simple velocity by depth graph where there are areas of high risk to the survivability of the submarine in case of a hydroplane jam or flooding, this is very well explained on Martin Renilson's book [6].

## 3. MATHEMATICAL MODELLING

Submarine modeling is based on hydrodynamic equations of bodies in fluids. In the horizontal plane, the model of the TPN Simulator for surface vessels was maintained (Tannuri et al., 2014 [10]).

The motion and stability in the vertical plane depend on static factors, such as the hydrostatic restoring moment in pitch ($mg\overline{BG}.\theta$), and on hydrodynamic effects. To model the motion on the vertical plane, the Spencer (1968) [9] equations were initially adopted. They depend on only linear hydrodynamic derivatives, according to equations (1) and (2):

*Vertical plane Equations (Spencer)*

$$(m - Z_{\dot{w}})\dot{w} - wZ_w - (mU + Z_q)q - wZ_q = Z(t) \quad (1)$$

$$(I_y - M_{\dot{q}})\dot{q} - qM_q - wM_w + mg\overline{BG}.\theta = M(t) \quad (2)$$

The equations represent the movement of heave and pitch, respectively. In this way, cross-couplings between vertical and horizontal planes are disregarded. Such equations were not able to represent the submarine's behavior under different speed ranges and application, because of the

small angles linearization hypothesis. Spencer's model is only suitable for the speed range in which the hydrodynamic derivatives were obtained, a natural limitation of a linear model that truncates the other terms of the Taylor expansion.
Therefore, a more complex model is necessary to represent the submarine's behavior in all operating speed ranges. Feldman's equations (1979), which better represent the hydrodynamic forces and moments, were adopted. The numerical model behaved in a manner consistent with a real submarine in all conditions, which could be verified by a qualitative comparison between simulation results and tests on real submarines. Feldman's equation for heave and pitch motion (vertical plane) are described below on equations (3) and (4), notations are in references (Feldman, 1979) [2].

*Normal Force Equation (Feldman)*

$$
\begin{aligned}
&m[\dot{w} - uq + vp - z_G(p^2 + q^2) + x_G(rp - \dot{q}) \\
&\qquad + y_G(rq + \dot{p})] = \\
&+\frac{\rho}{2} l^4 Z'_{\dot{q}} \dot{q} \\
&+\frac{\rho}{2} l^3 [Z'_{\dot{w}}\, \dot{w} + Z'_q\, uq + Z'_{vp}\, vp] \\
&+\frac{\rho}{2} l^2 [Z'_*\, u^2 + Z'_w\, uw] \\
&+\frac{\rho}{2} l^2 \left[Z'_{|w|}\, u|w| + Z'_{ww} \left|w(v^2 + w^2)^{1/2}\right|\right] \\
&+\frac{\rho}{2} l^2 \left[Z'_{\delta s}\, u^2\delta_s + Z'_{\delta b}\, u^2\delta_b + Z'_{\delta s\eta}\, u^2\delta_s \left(\eta - \frac{1}{C}\right) C\right] \\
&-\frac{\rho}{2} c_d \int_{l_i}^{l_f} b(x)\; w(x)\, \{[w(x)]^2 [v(x)]^2\}^{1/2} dx \\
&+\frac{\rho}{2} l\, \bar{c}_L \int_{x_2}^{x_1} v(x)\, \bar{v}_{FW}(t - \tau[x]) dx \qquad (3)
\end{aligned}
$$

*Pitching Moment Equation (Feldman)*

$$
\begin{aligned}
&I_y\dot{q} + (I_x - I_z)rp - (\dot{p} + qr)I_{xy} + (p^2 - r^2)I_{zx} + (qp \\
&\qquad - \dot{r})I_{yz} \\
&+m[z_G(\dot{u} - vr + wq) - x_G(\dot{w} - uq + vp)] = \\
&+\frac{\rho}{2} l^5 [M'_{\dot{q}}\dot{q} + M'_{rp} rp] \\
&+\frac{\rho}{2} l^4 [M'_{\dot{w}}\dot{w} + M'_q uq] \\
&+\frac{\rho}{2} l^3 [M'_* u^2 + M'_w uw + M'_{w|w|r} w|(v^2 + w^2)^{1/2}|] \\
&+\frac{\rho}{2} l^3 [M'_{|w|} u|w| + M'_{ww} |w(v^2 + w^2)^{1/2}|] \\
&+\frac{\rho}{2} l^3 \left[M'_{\delta s} u^2\delta_s + M'_{\delta b} u^2\delta_b + M'_{\delta s\eta} u^2\delta_s \left(\eta - \frac{1}{C}\right) C\right] \\
&+\frac{\rho}{2} C_d \int_{li}^{lf} x\, b(x)\, w(x)\, \{[w(x)]^2 + [v(x)]^2\}^{1/2}\, dx \\
&-\frac{\rho}{2} l\bar{C}_L \int_{x_2}^{x_1} x\, v(x)\, \overline{v}_{FW}(t - \tau[x])\, dx \\
&\qquad -(x_G W - x_B B) cos\theta\; cos\phi - (z_G W - z_B B) sin\theta \; (4)
\end{aligned}
$$

The simulation software had to undergo a series of adaptations at the programming level. An example was the horizontal control surfaces, called hydroplanes, which do not exist on surface ships. These actuators were modeled and implemented using Matlab® functions integrated in real-time with the maneuvering simulator core (C++).
The simulator can represent the submarine navigation on the surface and underwater. When navigating on the surface (see the submarine in Guanaraba Bay in Figure 2), the mathematical model of the surface ship is adopted, with minor modifications. For example, the submarine rudders do not operate on the propeller's wake such as in a ship. When the submarine is submerged, the modified model based on Feldman's equations is applied. Further details of the mathematical models can be found in Nascimento (2019) [5].

Figure 2: Simulator, the submarine operating on surface, and below the submarine diving (source: TPN-USP)

## 4. THE SUBMARINE MANEUVER MODEL

The numerical model was created to represent an 80m length submarine, like a nuclear submarine. The hull geometry adopted was the DARPA Suboff hull ([4] Grooves 1989) since it is a hull with unclassified data and suitable for academic work. The maximum design speed was 25 knots, and the propeller characteristics were inspired by the one used by the Brazilian Navy in its submarine free-running Model Number 02 (Sbragio, 2018 [8]). The control surfaces (rudders and hydroplanes) were also the same as for DARPA. The main characteristics are described as follows:

Table 1: Main Submarine Particulars

| | |
|---|---|
| LOA | 82,3 m |
| LBP | 80,5 m |
| Hull Diameter | 9,60 m |
| Submerged Displacement(Δ) | 4890 ton |
| Maximum Speed | 25 knots |
| BG | 30 cm |
| 2Rudders+2 Hydroplanes | 8,85 m² each |
| Forward Hydroplanes | None |
| Propeller ML02-MB 7blades | FPP 5,66m |

The visual and graphics part was produced by the TPN-USP team. It includes the 3D model of the submarine, some operation panels (Figure 3), the underwater view (Figure 2 below), as well as the 3D model of the propeller and all other features necessary for the simulator.

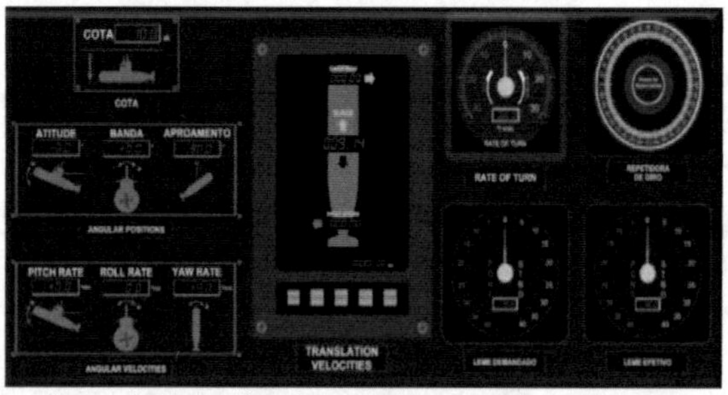


Figure 3: Example of Control and Equipment Panels (source: TPN-USP)

The hydrodynamic coefficients for fully submerged condition were obtained by tests carried out by the US Navy (Roddy, 1990 [7]) and also by the use of the CAE software of CFD StarCCM+ for both submerged and on surface condition, since there were no coefficients available for the model operating on the surface. The hull was worked on Rhinoceros. The numerical meshes had 3 to 20 million elements, depending on the refinement required for the simulation. Figure 4 show a simulation on Star-CCM+ to obtain hydrodynamic coefficients on the surface, this case required a better detailing of the mesh on the free surface and had approximately 19 million elements. Table 2 presents the Force and Coefficient with 5° and 10° incidence angles from the bow in the horizontal plane with a flow of 12.9m/s, with a comparison between the CFD simulations and the coefficients obtained by Roddy (1990) [7]. The nondimensional factors are $\mathbf{0.5\rho LDuU}$ for force and $\mathbf{0.5\rho L^2DuU}$ for the moment. More details can be found in Nascimento (2019) [5]. Note that the difference was less than 10% for 5° and 10° incidence angles. CFD calculation was executed 2°, 5° and 10° incidence angles from the bow, and the results were used to calculate the linear derivatives $\boldsymbol{Y_v}$, $\boldsymbol{N_v}$, $\boldsymbol{Z_w}$ *e* $\boldsymbol{M_w}$.

There are results for 90° incidence angle from the bow in the horizontal plane (abeam incidence) of a uniform flow, showing that the CFD results are quite different from the calculated by the linear coefficients of DARPA. This is expected since for such large angles, the model of linear hydrodynamic derivatives is not valid anymore. Also, for better calibration of the model, simulations were performed with a uniform flow of 90° incidence angle from the bow in the vertical and horizontal plane, as shown in the pressure distribution for lateral flow in Figure 4 (horizontal case).

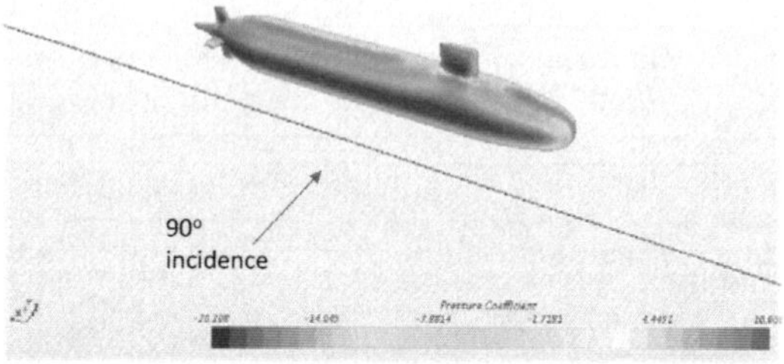


Figure 4: Pressure distribution for flow 90 ° to 0.1m/s

Table 2: Results by CFD for the horizontal plane and deep submerged compared with DARPA by Roddy(1990) [7]

| Method | Lateral Force (kN) | Moment (kNm) |
|---|---|---|
| Angle of Incidence | 5 | U (m/s) = 12.86 |
| DARPA | -1332.24 | -52582.95 |
| CFD | -1462.86 | -55300.00 |
| Coefic | -0.02034 | -0.00997 |
| Coef CFD | -0.02233 | -0.01045 |
| Angle of Incidence | 10 | U (m/s) = 12.91 |
| DARPA | -2664.67 | -105173.04 |
| CFD | -3494.61 | -108348.00 |
| Coef DARP | -0.04052 | -0.01995 |
| Coef CFD | -0.05315 | -0.02047 |
| Angle of Incidence | 90 | U (m/s) = 0.10 |
| DARPA | -0.92 | -36.48 |
| CFD | -2.09 | -2.14 |
| Coef DARP | -0.23337 | -0.11442 |
| Coef CFD | -0.52871 | -0.00671 |

An analysis of stability in the vertical plane was made using Feldman's equations of motion. An important peculiarity of submarines in the vertical plane is the presence of hydrostatic restoration that does not exist in the horizontal plane. The submarine may be directionally stable for low speeds, but unstable for moderate speeds. The hydrodynamics of the hull becomes dominant, and the submarine can become unstable (straight line instability), depending on the shape of the hull and rudders/hydroplanes area.

It was shown that for the DARPA SUBOFF hull in the original configuration (without nozzle) with rudders/hydroplanes of 8.85 m² each (in the full scale used for this work) is unstable, however with rudders/hydroplanes with area of 12 m² they are already stable at any speed. This vertical stability is highly desirable in any real submarine project, since a hydroplane jam in a neutral position brings the submarine to trim angle of zero rather than keep increasing the trim as in a unstable case.

The stability can be observed by the indexes, as shown in Figure 6. It contains the nondimensional roots of the characteristic equation (5) (Roddy,1990) [7]. The real part of the indexes shown in green and the complex ones in dashed red.
The submarine is very directionally stable up to 5 knots, reducing the stability margin at approximately 10 knots and becoming unstable at speeds above 10 knots (original control surfaces size).

*Laplace form characteristic equation (Roddy [7])*

$$[(Z'_{\dot{w}} - m')(M'_{\dot{q}} - I'_y) - (Z'_{\dot{q}} + x'_G m')(M_{\dot{w}} + x'_G m')]s^3$$
$$+[(Z'_{\dot{w}} - m')(M'_q - x'_G m') + Z'_w(M'_{\dot{q}} - I'_y)$$
$$-(Z'_q + x'_G m')M'_{\dot{w}} - (Z'_{\dot{q}} + m')(M'_{\dot{w}} + x'_G m')]s^2$$
$$[Z'_w(M'_q - x'_G m') + (Z'_{\dot{w}} - m')M'_\theta - (Z'_q)M'_w]s$$
$$+Z'_w M'_\theta = 0 \qquad (5)$$

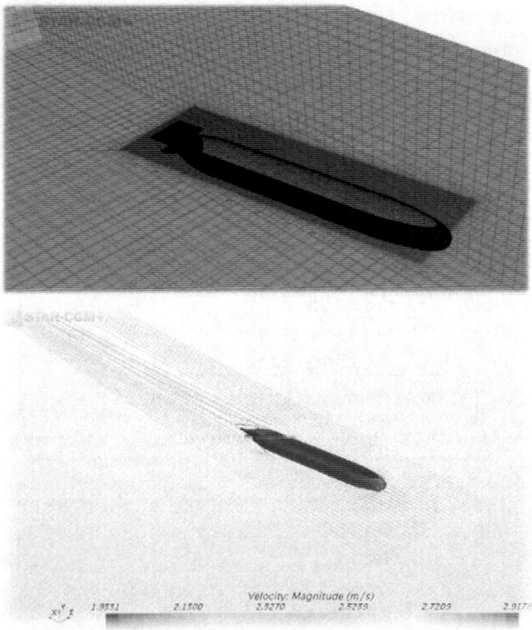


Figure 5: CFD simulation on Star-CCM + to obtain hydrodynamic coefficients on the surface. Above, the mesh with 19 million elements in a simulation of 10° from starboard, and below, the flow streamlines.

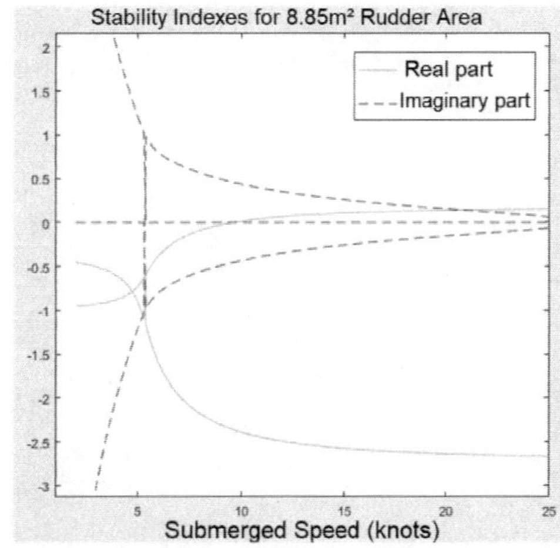


Figure 6: Stability indexes for movement in the vertical plane as a function of speed

## 5. MAIN RESULTS

An interface to test the submarine maneuver was developed, running in "Fast-Time". In this way, this interface makes it possible to analyze the submarine's maneuvering characteristics much faster and make the analysis or design process much more efficient. An autopilot for heading and depth control was also implemented with PD control law, so the operation of the submarine is much simpler and more reliable and allows the analysis of control systems and the effectiveness of the rudders/hydroplanes.
In addition to a model to operate submerged, which was the focus of the work, a submarine model was also created for surface operation, for port navigation. Thus, it is possible to simulate port inbound and outbound manoeuvres, considering the use of tugs and environmental factors, including in the locations of Itaguaí-RJ and Guanabara Bay, where there are submarine bases of the Brazilian Navy.
From the model it was possible to perform several maneuvers in the horizontal and vertical plane:
- Turning circle and pullout;
- Zigzag in the horizontal plane for different rudder angles;
- Direct spiral and Bech spiral;
- Zigzag in the vertical plane; and
- Meander in the vertical plane.

Next, in Figure 7, we can see the result of a Turning Circle performed with the initial speed of approximately 12 knots and submerged and below a 10°/10° Zigzag in the horizontal plane. Each maneuver performed was carried out at different speeds and conditions.

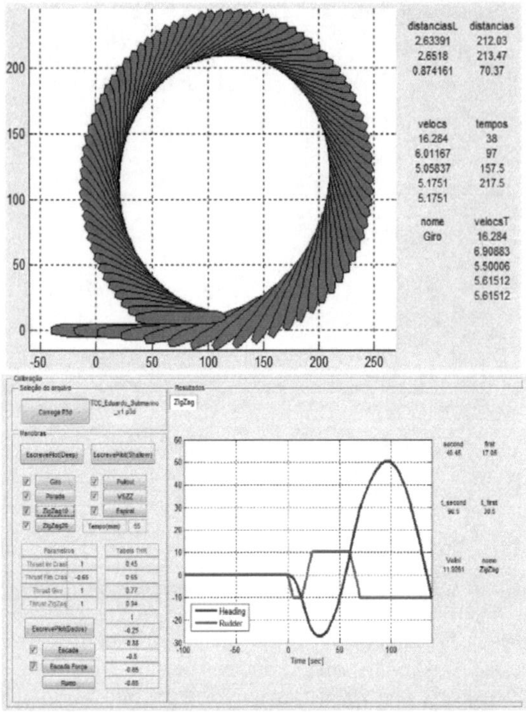


Figure 7: Interface for automated execution of manoeuvre in Fast-Time, showing the turning circle and below the results of a 10°/10° Zigzag in the horizontal plane

Several maneuvers were also carried out in the vertical plane, which are even more important for the operation of submarines. Some results are presented in the following figures for a maneuver of depth change at 6 knots using 20° with the aft hydroplane. Figure 8 shows the depth, trim angle and pitch angle speed versus time, respectively.

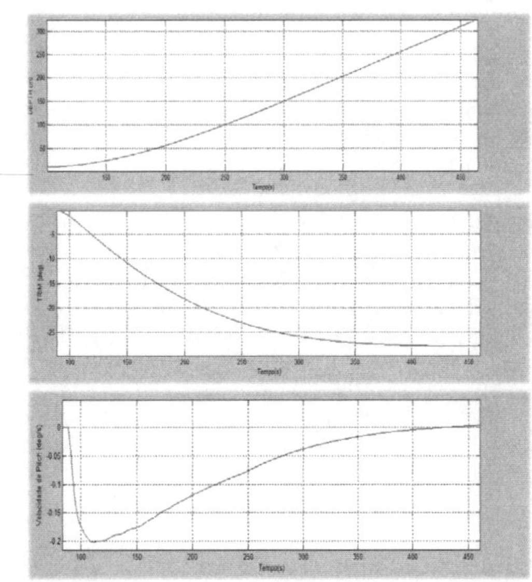

Figure 8: From top to bottom: depth, trim angle and pitch velocity versus time in a depth change manoeuvre

It can be noted that after a few minutes the submarine stabilizes in a sharp 27° downward trim.

Another important result related to control is the Zigzag in the vertical plane. Among the various tests carried out, in Figure 9 it is presented with the submarine at 6 knots and using 7.5 ° of angle in the aft hydroplanes to execute a vertical plane Zigzag. It can be observed a large overshoot in depth (blue dashed line) in the same way the horizontal plane Zigzag maneuver presents an overshoot of heading. Table 3 condense the main manoeuvre results.

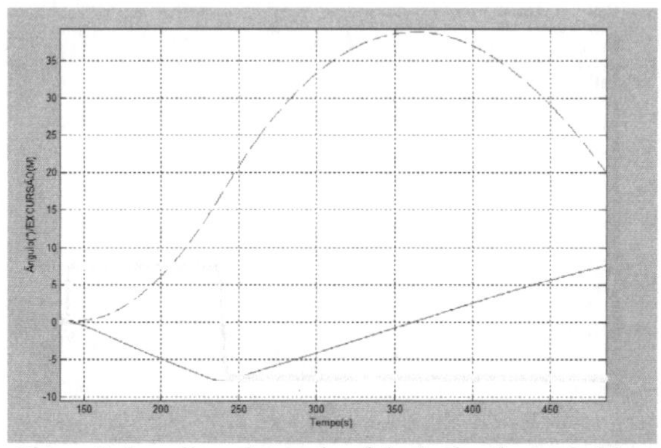


Figure 9: In yellow the angle of the hydroplanes, in blue the full line the pitch speed and in blue dashed the depth in meters

Table 3: Main Maneuver Results

| | |
|---|---|
| Submerged tactical diameter | 212 m |
| Advance 90° of heading change | 223m |
| Transfer 90° of heading change | 72 m |
| Overshoot Zigzag 10° at 12 knots | 27° |
| Overshoot of depth Zigzag 7.5° at 6 knots | 39,5 m |
| Critical velocity for aft hydroplane | 3.5 knots |
| Critical velocity for vertical dynamic instability (should be very high in a real submarine project) | 10 knots |

## 6. CONCLUSIONS

This paper summarizes the development of the TPN-USP Submarine Maneuvering Simulator. It was based on the existing TPN Ship Maneuvering Simulator when operating on the surface, integrated to the Feldman's equations (1979) when submerged. In figure 10 is shown the submarine operating at surface in a simulation of navigation to dock in the Brazilian naval submarine base. We concluded that the well-known equations for submarine maneuvering (such as Spencer and Feldman's equations) are not well suitable to represent the submarine behaviour for simulation purposes. They are enough to predict some basic manoeuvres such as turning circles and Zig-zag, but are not suitable for high incidence angles performance or complex emergency manoeuvres. Therefore, for horizontal plane was kept the existing numerical model of TPN manoeuvring simulator and a

 

simplified version of Feldman's equations was adopted for the vertical plane.
Other relevant fact is that linear equations proposed by Spencer are enough to predict the stability at a determined speed, but not to represent the correct hydrodynamic behavior of the submarine through the whole range of speeds if proper corrections are not implemented.
CFD simulations for larger incidence angles (e.g. 20°) demonstrated a strong pitching moment tending to bow up the ship due a large area of separation flow. This phenomenon is useful in emergency recovery maneuvers when the submarine needs to come to the surface quickly. Such emergency manoeuvres can be used to design the Safety Operational Envelope as stated by Renilsion [6].
Finally, we can think of possibilities of improvement for work or others that would use this work as a starting point/tool. Some examples of future possibilities are:

- To merge the surface model with the submerged one and include the hydrodynamics close to the surface to represent the simulation in any situation and to improve the equation of the numerical model;
- Inclusion of the dynamic system of Ballast Tanks and Compensation Tanks to assess the influence on the maneuver;
- Calibration to represent existing Submarines or creation of a Digital Twins;
- Risk analysis in several areas (maneuverability, port maneuver);
- Improvement to assess the influence of environmental factors (surface and submerged);
- Determine Safe Operating Envelopes from a calibrated model with data of model testes (captive or free model);
- Autopilot development; and
- Use the Submarine simulator as an environment for testing "guided" Torpedo algorithms.

Figure 10: Entry/Exit of a Submarine from the Itaguaí Brazilian Submarine Base (source: TPN-USP)

## 7. ACKNOWLEDGEMENTS

The authors acknowledge the Brazilian Navy for providing the data supporting this development and all the members of TPN-USP envolved. The second author also thanks the CNPq (process 310127/2020-3) for the research grant.

## 8. REFERENCES

1. ARENTZEN, E. S. AND MANDEL, P. E., 'Naval Architectural Aspects of Submarine Design', *Trans. SNAME, pp. 622-692*, 1960.

2. FELDMAN J. DTNSRDC, 'Revised Standard Submarine Equations of Motion SPD-0398-09' *David Taylor Research Center*, 1979.

3. FOSSEN T. I., 'Handbook of Marine Craft Hydrodynamics and Motion Control', *Trondheim, Norway*, 2011.

4. GROOVES, N.C.; HUANG, T.T.;CHANG, M.S., 'Geometric Characteristics of the DARPA SUBOFF Models (DTRC Models No. 5470 and 5471) DTRC/SHD-1298-01', *David Taylor Research Center, Bethesda, MD, USA*, 1989.

5. NASCIMENTO C.E.G., 'Manobra de Submarinos em Simuladores', *University of São Paulo, Brazil*, 2019.

6. RENILSON M., 'Submarine Hydrodynamics', *Launceston, TAS, Australia*, 2015.

7. RODDY R.F. 'Investigation of the Stability and Control Characteristics of Several Configurations of the Darpa Suboff Model (DTRC Models No. 5470) AD-A227715', *David Taylor Research Center, Bethesda, MD, USA*, 1990.

8. SBRAGIO R. ET AL, 'Development of a Propeller for an Autonomous Underwater Vehicle with a hull geometry of the Darpa Suboff Model', *LABHIDRO, CONEM, Brazil*, 2018.

9. SPENCER J.B, 'Stability and Control of Submarines', *Ship Department, Ministry of Defense, USA. Reprint of the Royal Naval Scientific Service, vol. 23., pp. 187-345*, 1968.

10. TANNURI, E. A.; RATEIRO, F. ; FUCATU, CARLOS H. ; FERREIRA, MARCOS D. ; MASETTI, I. Q. ; NISHIMOTO, K., 'Modular Mathematical Model for a Low-Speed Maneuvering Simulator', *ASME 33th International Conference on Ocean, Offshore and Arctic Engineering OMAE 2014, San Francisco*, 2014.

11. TANNURI, E. A.; MARTINS, G. H. A., 'Application of a maneuvering simulation center and pilots expertise to the design of new ports and terminals and infrastructure optimization in Brazil', *34th PIANC World Congress 2018, Panama City*, 2018.

## 9. AUTHORS BIOGRAPHY

**Carlos Eduardo Guedes do Nascimento** hold the current position of Head of Modeling Division at the Naval Architecture Department of the Submarine Development Center at the Brazilian Navy. He is responsible for 3D modeling and for the hydrodynamic, maneuverability and seakeeping of the nuclear Brazilian submarine project. His previous experience was onboard Brazilian Navy warships as Officer of the Watch, navigation and machinery Officer. Bachelor in Naval Sciences (Brazilian Navy Naval Academy, 2014) and Naval and Oceanic Engineering (Escola Politécnica da Universidade de São Paulo, 2020).

**Eduardo Aoun Tannuri** received the degree in mechatronics engineering and the Ph.D. degree in control from the Universidade de São Paulo (USP), Brazil, in 1998 and 2003, respectively. He is currently a Full Professor with the Department of Mechatronics Engineering, USP and director of USP Ship Manoeuvering Simulation Center. He has published 195 papers in conferences and 45 articles in journals. His research interests include non-linear dynamics and control, with applications in autonomous maritime vehicles and ship maneuverability. He has been a member of the International Towing Tank Conference (ITTC) Maneuvering Committee, since 2011. He received the Best Student Paper Award in the IFAC CAMS Conference in 2001 and the National Prizes (the Brazilian Navy Engineering Merit Award in 2017 and the National Agency for Petroleum, Natural Gas and Biofuels Innovation Prize in 2019).

# DESIGN OF DELTA WING AUTONOMOUS UNDERWATER GLIDERS FOR LONG TERM SUSTAINABLE AND STEALTH RECONNAISSANCE USING NUMERICAL STUDIES

**M Guggilla** and **R Vijayakumar** Indian Institute of Technology Madras, India

## SUMMARY

Autonomous Underwater Gliders (AUGs) are a unique source of collecting oceanographic data for varying depths with longer endurance than other types of unmanned underwater vehicles. The buoyancy-driven propulsion mechanism assists in diving deep into the ocean, collect data, resurface in saw-tooth and spiral motions. There have been various hull-forms introduced for this purpose starting from - Torpedo hull, XRay Flying-Wing Glider, etc., all of which have seen an improved understanding of the hydrodynamics associated with the hull-form of the gliders. The saw-tooth motion represents the glider's capacity to cover larger distances with a smaller angle of attack, thereby achieving an increased range for the same power included within the hull batteries. In this paper, a design study of delta wing hull-forms, which are proven to have better longitudinal motion characteristics, is proposed to arrive at a preliminary hull form that can be used for long term reconnaissance missions, remaining in stealth to assist submarines with valuable data. Delta wing hull forms are varied using different NACA sections, and numerical analysis of the obtained hull forms is performed to arrive at the hydrodynamics; namely, the drag, lift forces and coefficients for varying angles of attack, using commercial CFD software – StarCCM+. These results are then compared against each other to derive the preliminary hull form for further study on better longitudinal motion characteristics, thereby using less power for its motion underwater.

## NOMENCLATURE

| | |
|---|---|
| $\alpha$ | Angle of Attack (degree) |
| $\upsilon$ | Kinematic viscosity (N s m$^{-2}$) |
| $\rho$ | Density of water (kg m$^{-3}$) |
| $L$ | Lift force (N) |
| $D$ | Drag force (N) |
| $L/D$ | Lift-to-Drag Ratio |
| $C_L$ | Coefficient of Lift |
| $C_D$ | Coefficient of Drag |

## 1. INTRODUCTION

The Underwater Gliders are a unique set of oceanographic data collection mechanisms in use since the late 1900s. Henry Stommel [1] envisioned utilising a buoyancy-controlled device with wings that can be used to map the data underwater vis-a-vis the temperature, salinity, density, and such in 1989. These platforms can glide through the ocean in alternating cycles of diving and surfacing and arrive at the desired location to spiral down to a given depth and sample the water column. This approach inspired a generation of glider models and prototypes, including the Slocum [2], Spray [3] and Seaglider [4] models, which are traditionally called the legacy gliders. Jenkins et al., [5], in their work, have given a classification of gliders for the different roles/mission requirements that it has to perform. These roles include deep zig-zag motion, flat glide sloped depth-limited movement, virtual station keeping, etc.

The gliders perform two significant motions during their operation, the zig-zag or saw-tooth motion and the spiral motion. The saw-tooth motion is the typical diving and surfacing motion repeated as a cycle. The main wings, the moving mass and the controllable bladders provided inside the model work in tandem to execute the zig-zag motion. The spiral motion is also largely dependent on the stability of the glider ensured by the roll and pitch control masses. These motions are performed with minimal or no noise keeping the ambient ocean conditions constant for better sampling and observation. This is primarily due to the absence of traditional propulsion devices in the design. Due to this inherent nature, autonomous underwater gliders tend to be very stealthy and therefore result in better reconnaissance missions. In addition, the batteries included can power the gliders for months without the need to service or repair them. Many gliders have circumnavigated the globe over several months and have thus proven long term sustainability.

Other areas of glider operations include monitoring the marine life, coastal zone and the minerals available such as manganese nodules, etc. All these missions require different design conditions, including but not limited to the speed, area of operation, maximum dimensions, buoyancy required, and payload. For a task requiring higher forward speed, the traditional legacy gliders derived from Myring hull designs [6] may not be fruitful. But a glider focussed on high lift characteristics will prove to be a better candidate due to its inherent advantages. Still, the span of such gliders is much larger than the wings of a torpedo hull, thereby limiting them in certain conditions of depth. Examples of these gliders include the Liberdade class X-Ray [7] and Z-Ray [8] gliders. Comparison of lift and drag forces between a legacy glider model and a delta wing model of equal volume shows that the delta wing models fare much better for longer endurance missions. Comparing the delta wing glider models with a classical Myring hull design has also been studied to understand the performance in terms of L/D better [9]. In this paper, a design study of a delta wing underwater glider is shown. Three different NACA Sections are used for the delta wing designs and the resulting glider models are numerically simulated for different speeds in water to obtain the hydrodynamic forces on each glider model. The hydrodynamic coefficients of lift ($C_L$) and drag ($C_D$) are taken, and the lift-to-drag ratios (L/D) are plotted to show their variation as the models are varied.

## 2. GLIDER DESIGN

The initial design of a hull form requires the mission specific requirements. The available information during the start of this study included the maximum wingspan that can be used, the capacity of the buoyancy engine, the minimum and maximum forward speeds required and the estimated pressure hull volume. The pressure hull houses the equipment required for the glider operation, which include a charge-coupled camera sensor, Obstacle Avoidance sensor, Conductivity-Temperature-Depth Sensor, Inertial Navigation System, Sonars, Buoyancy Engine, Roll and Pitch control, on-board computer and battery pack. Three NACA sections – NACA0015, NACA0018 and NACA0021 are the primary wing sections that are blended individually for a span of 3 metres. The root chord length is taken as half of the total span of the glider. All the three models are designed to house a given volume for the equipment on board. The Models-1, 2 and 3 are designed envisioning a minimum forward speed of 0.5 m/s. As the cross-section of the main wings is varied across the models, the aft portion of the gliders, i.e. the trailing edge of the wing moves closer to the leading edge across the span. This results in differences in wake flow pattern at larger speeds.

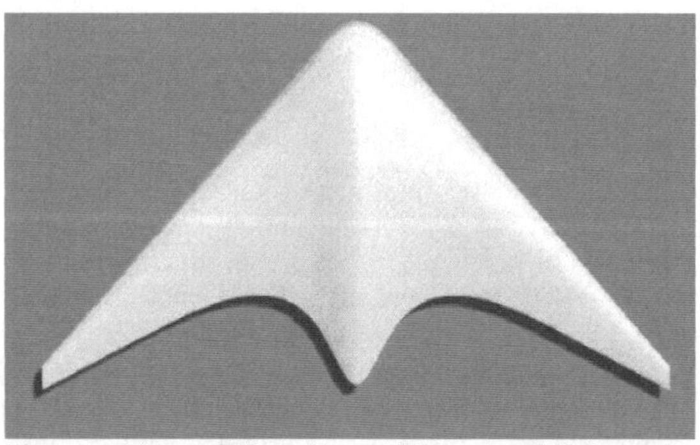

Figure 1(a): Delta-wing glider modelled using NACA0015 cross-section (Model-1)

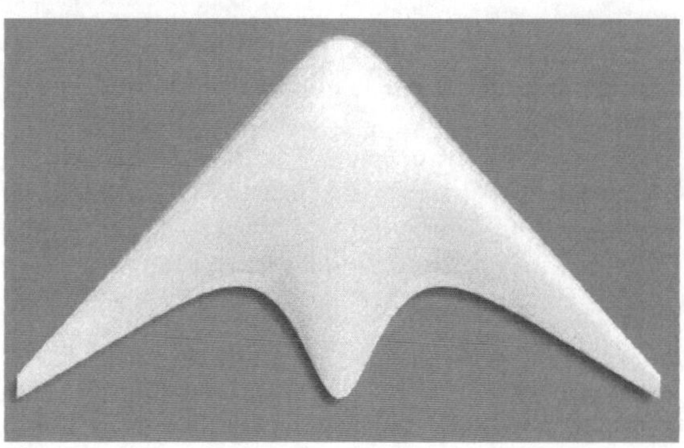

Figure 1(b): Delta-wing glider modelled using NACA0018 cross-section (Model-2)

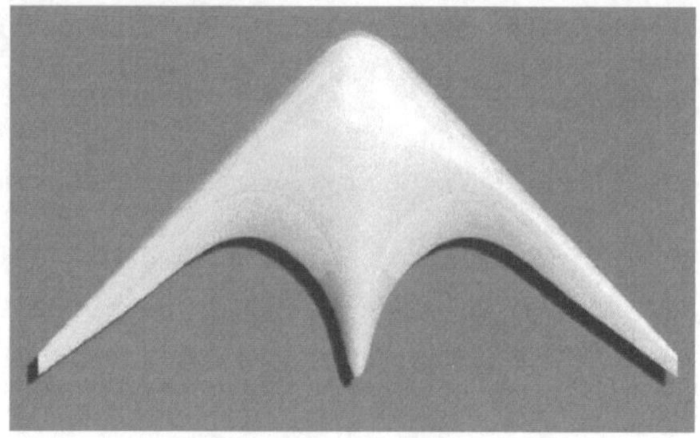

Figure 1(c): Delta-wing glider modelled using NACA0021 cross-section (Model-3)

The streamlined hull forms provide a favourable pressure gradient over the cross-section thereby assisting the low-drag designs to achieve significantly higher L/D. The resultant hull models are shown in the Figure.1. The 3 models are designed in Rhinoceros and imported into Star-CCM+ to numerically simulate the flow past the glider models and obtain the hydrodynamic forces acting on them for the given flow velocities of 0.4 m/s, 0.8 m/s and 1.2 m/s which is the expected speed range for the model. For each velocity condition, the angle of attack of the glider is changed from $0^o$ to $12^o$.

## 3. NUMERICAL MODELLING

The flow past a glider can be simulated using rigid body motion equations [10] and many numerical models [11] have been developed for the same. The current study uses the commercial solver, STARCCM+, for the numerical modelling of the glider models-1, 2 and 3. The domain created for the flow is a cuboid and as per the ITTC specifications [12], as shown in the Fig.2 below. The forward boundary is placed at a distance of 1.3 times the total root chord length of the glider, the velocity of the flow is specified at this boundary. The boundary to the aft of the glider is taken as a pressure outlet and is at a distance of 5 times the chord from the aft. The remaining radially outward faces are placed at 1.3 times the root chord length.

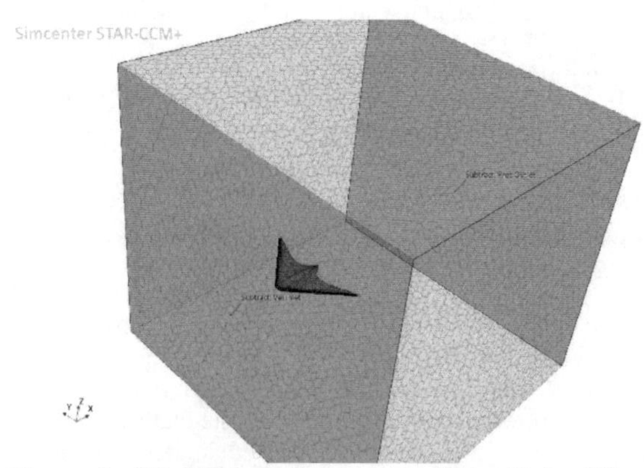


Figure 2: 3D Meshed Domain used for Delta Wing Underwater Glider

The density is kept constant throughout the domain and a steady flow solver is used for the simulations. The SST k-ω turbulence model is adopted for the flow due to its better performance in flow visualization at the aft in the wake region. The grid used in the domain is unstructured in nature as shown in Figure 3(a). This polyhedral meshing has performed consistently better in achieving faster convergence when compared to structured meshing, thus reducing the computational costs. The polyhedral meshing applied to the glider models is shown in Figure 3(b).

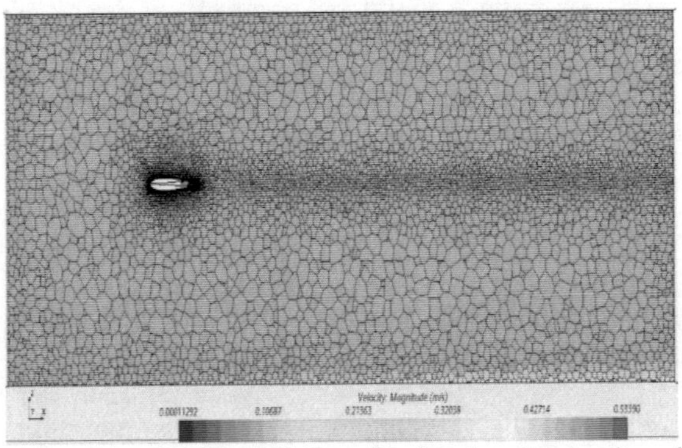

Figure 3(a): 2D Meshed Domain showing the density of cells inside the domain

Figure 3(b): 3D Meshed Model of the NACA0018 Delta Wing hull form (Model-2)

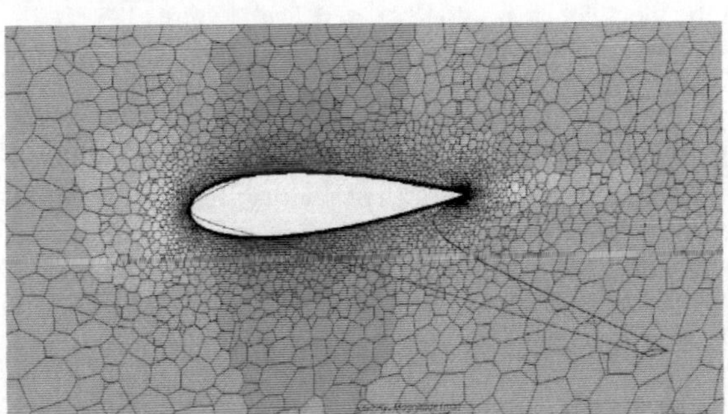

Figure 3(c): Cross-section showing the polyhedral meshing around the surface of the model

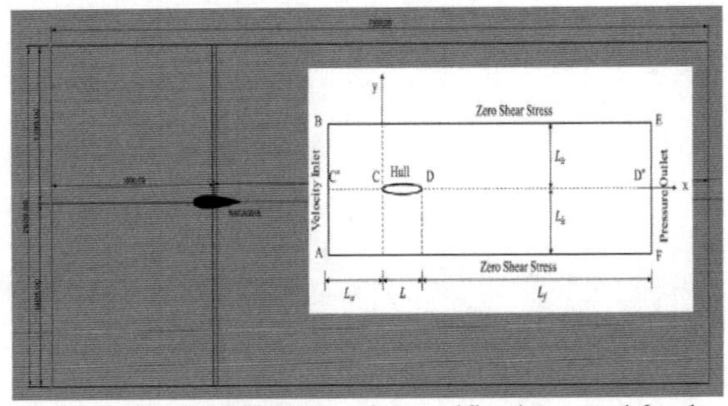


Figure 3(d): The ITTC Domain specifications used for the numerical model's domain

The surface meshing on the NACA0018 glider as shown has greatly reduced the number of cells required to accurately mesh the shape of the glider models. The y-plus value is set at one for the turbulence model with a turbulence intensity of 10. The surface of the glider model is additionally meshed as a wall with prism layers for a more accurate flow analysis (Figure 3(c)). The velocity inlet and pressure outlets are at the forward and aft of the glider respectively. The remaining sides are modelled as walls with no-slip boundary conditions (Figure 3(d)). The Kinematic viscosity, $\upsilon$ is taken as 8.8871e-7 $m^2\ s^{-1}$ and the density of water, $\rho$ is taken a 1000 kg $m^{-3}$

The grid independency study is taken up as the next step in the numerical process. The Drag force obtained for the Model-2 as mesh is refined and the number of cells are increased is plotted as shown in Figure 4. The graph shows that the convergence of the drag force starts at approximately 4.2 million cells. The flow velocity observed is 0.8 m/s and the angle of attack of the flow is zero.

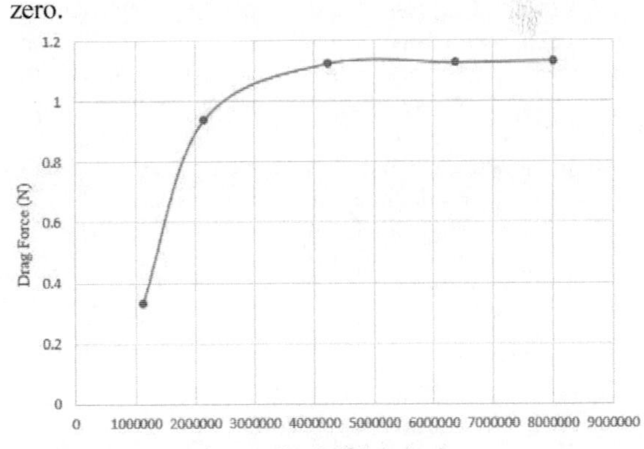


Figure 4: Grid Independency test showing the convergence of drag

## 3.1 NUMERICAL MODEL VALIDATION

The numerical model generated using the above domain and boundary conditions is validated using the results from Sun et al. [13]. A NACA0012 cross section is taken with a chord length of 444 mm and a span of 500 mm.

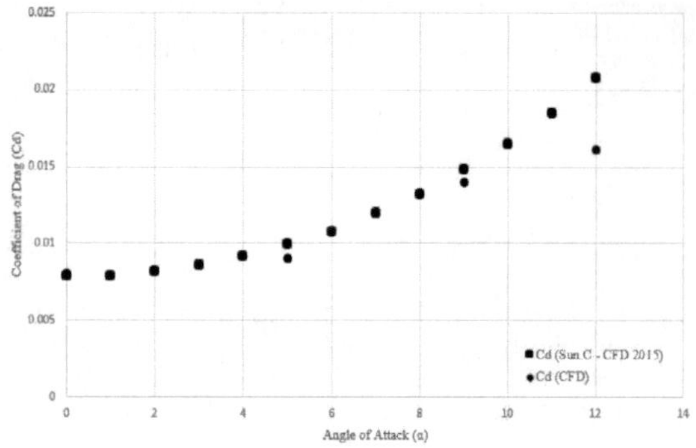


Figure 5: Comparison of Coefficient of Drag for a NACA0012 section between Sun et al and the current numerical model

The angle of attack of the flow with a velocity of 0.4 m/s, is varied from $0^o$ to $12^o$ and the coefficient of the drag force obtained is used for the model validation. The average error is approximately 10% and therefore within the desired range. As shown in Figure 5, the drag predicted in the case of the numerical model being validated compares very well at lower angles of attack but as the angle of attack increases, the drag is over predicted by the numerical model due to the presence of fully developed turbulence across the cross section of the aerofoil.

## 4. RESULTS

The validated numerical model is used to simulate the flow over the glider models-1, 2 and 3. The hydrodynamic forces, lift and drag, are obtained for flow velocities of 0.4 m/s, 0.8 m/s and 1.2 m/s. The corresponding length-based Reynolds number range is 600,000 to 2,000,000. The simulations are all performed under fully turbulent conditions. The Angle of Attack for each velocity goes from 0° to 12° at intervals of 3°. The following graphs shown in Figure 6(a, b and c), show the L/D plots for each of the above flow velocities respectively.

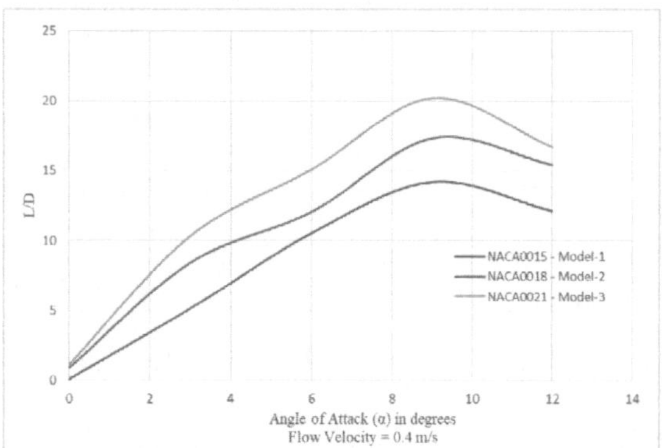


Figure 6(a): Comparison of Lift-to-Drag Ratios of the Delta Wing Models-1, 2 and 3 for a flow velocity of 0.4 m/s.

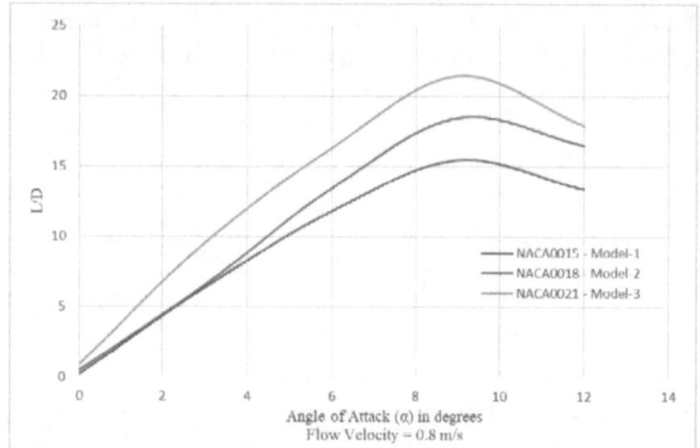


Figure 6(b): Comparison of Lift-to-Drag Ratios of the Delta Wing Models-1, 2 and 3 for a flow velocity of 0.8 m/s.

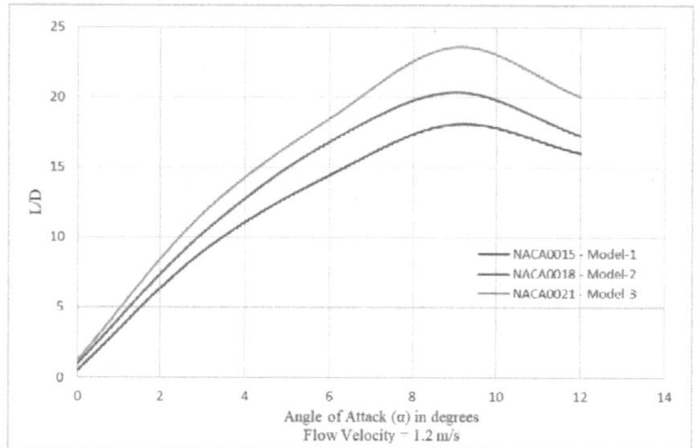


Figure 6(c): Comparison of Lift-to-Drag Ratios of the Delta Wing Models-1, 2 and 3 for a flow velocity of 1.2 m/s.

For all the three flow velocities, the models have varying L/D ratios but generally following the trend with respect to a given range of angles of attack. The monotonous increase in the lift-to-drag ratio for a given hull model is largely due to the aerofoil cross section used to model the delta wing glider. The stall phenomenon observed in all the graphs occurs after 9° angle of attack. The results provided do not include angle wise analysis from 0 to 12, but we can say that the maximum value of L/D occurs around the 9° angle with an error margin of 1°. For Model-1, as the velocity is varied from 0.4 to 1.2 m/s, there is an average increase of L/D at 14.3%. Similarly, the case for Model-2 and 3 at 16.15% and 8.78% respectively.

The Models have shown to have a higher lift and lower drag despite their large surface areas due to the characteristics of the wing design where in the pressure gradient is favourable till the flow reaches the maximum cross section of the wing. At higher speeds, the transition of the flow happens much faster leading to a larger turbulent boundary layer and wake regions, thus increasing the drag and reduction in the L/D ratio. For an angle of attack of 6° and at maximum speed, the models-2 and 3 have better lift characteristics and reach the desired L/D range between 15 and 20. This angle of attack is chosen for comparison on the basis of the characteristics of the two-dimensional cross section of the wing maximizing the performance. Model-2 delivers larger overall efficiency due to the presence of a smaller turbulence region compared to Model-3. But the target range of L/D is observed consistently for Model-3 for all the three speeds. The maximum L/D ratio for the glider model is approximately 20 which should be more than sufficient to achieve the target maximum speed. A concurrent study has been conducted on the effects of the characteristic wing parameters on the glider model for increasing the hydrodynamic efficiency [14]. The final decision on which glider model can be used may be arrived at by introducing the rudder and studying its effects and then plotting the saw-tooth motion of the glider models which is the future continuation of this study. The current glider designs can achieve the desired glide angle in between 20° and 30° on par with the existing gliders.

## 5. CONCLUSIONS

The delta wing glider design was taken up which is quite a challenge in the naval architecture perspective. The preliminary design was majorly governed by the payload inside the pressure hull and the general arrangement. Once the maximum size was estimated, for the target range of L/D, three NACA sections were chosen. These profiles acted as the cross-sections for building the models-1, 2 and 3 corresponding to NACA0015, NACA0018 and NACA0021 respectively. All the three have approximately equal volume for housing the necessary equipment. The designs were modelled in Rhinoceros and imported into the numerical solver in Star-CCM+. The domain and meshing were carried out as per ITTC specifications and validated using the published results. The flow over the models was simulated using the SST k-ω turbulence equations which provided for better wake visualization due to the models being largely varied at the trailing edge of the wings. The hydrodynamic drag and lift was obtained and then the L/D ratios were calculated. The plots show an increasing L/D for increasing angle of attack for an overall average of approximately 13% across the velocities and the three design models. Models-2 and 3 have shown to be better performers in terms of the range of L/D, flow separation and wake characteristics. Further analysis on their performance needs to be estimated using the appendages and trajectory simulation. But, based on the current data in this paper, it can be reasonably stated that the Model-3 is a better candidate for the next stage of design which is a part of future study.

## 6. ACKNOWLEDGEMENTS

This research is supported by the Department of Ocean Engineering, Indian Institute of Technology Madras under the Prime Minister's Research Fellowship programme. The authors also acknowledge the valuable inputs provided by Shashank Shankar, IIT Delhi.

## 7. REFERENCES

1. STOMMEL, H., 'The Slocum Mission', *Oceanography*, Apr 1989.

2. Webb, D.C., Simonetti, P.J., Jones, C.P., 'Slocum: an underwater glider propelled by environmental energy', *IEEE Journal of Oceanic Engineering*, 2001, Vol. 26, pp. 447-452.

3. Sherman, J., Davis, S.E., Owens, W.B., and Valdes, J., 'The autonomous underwater glider 'spray',' *IEEE Journal of Oceanic Engineering*, 2001, Vol. 26, pp. 437-446.

4. Eriksen, C.C., Osse, T.J., Light, R.D., Wen, T., Lehman, T.W., Sabin, P.L., Ballard, J.W., and Chiodi, A.M., 'Seaglider: A long-range autonomous underwater vehicle for oceanographic research', *IEEE Journal of Oceanic Engineering*, 2001, Vol. 26, pp. 424-436.

5. Jenkins, S.A., Humphreys, D.E., Sherman, J., Osse, J., Jones, C., Leonard, N., Graver, J., Bachmayer, R., Clem, T., Carroll, P., Davis, P., Berry, J., Worley, P., and Wasyl, J., 'Underwater Glider System Study', *Technical Report 53, Scripps Institution of Oceanography*, 2003.

6. Myring, D.F., 'A theoretical study of the effects of body shape and Mach number on the drag of bodies of revolution in subcritical axisymmetric flow', *Royal Aircraft Establishment, Farnborough, U.K., Tech. Rep. 81005,* 1981.

7. D'Spain, G.L., Jenkins, S.A., and Zimmerman, R., 'Underwater acoustic measurements with the Liberdade/X-Ray flying wing glider', *The Journal of the Acoustical Society of America*, Vol. 117(4), pp. 2624, 2005.

8. D'Spain, G.L., 'Flying Wing Autonomous Underwater Glider for Basic Research in Ocean Acoustics, Signal/Array Processing, Underwater Autonomous Vehicle Technology, Oceanography, Geophysics, and Marine Biological Studies', *Technical Report, Scripps Institution of Oceanography*, 2009.

9. Mukesh, G., Vijayakumar, R., 'CFD Study of the Hydrodynamic Characteristics of Blended Winged Unmanned Underwater Gliders', *International Society of Offshore and Polar Engineers, ISOPE*, 2019, pp. 1547-1552. ISBN 978-1 880653 85-2; ISSN 1098-6189.

10. Fossen, T.I., 'Guidance and Control of Ocean Vehicles', *4th Edition, Wiley & Sons*, 1994.

11. Shashank Shankar, R. V., and Vijayakumar, R., 'Effect of rudder and roll control mechanism

on path prediction of autonomous underwater gliders', *Lecture Notes in Civil Engineering*, 2019, Vol.22, pp 491-506.

12. Guggilla, M., and Vijayakumar, R., 'Study on the Hydrodynamic Performance of Unmanned Underwater Gliders with Varying Wing Sections Using CFD', *International Conference on Computational and Experimental Marine Hydrodynamics (MARHY)*, 2018, Pages 264-269, ISBN: 978-93-80689-30-2.

13. Sun, C., Song, B., and Wang, P., 'Parametric geometric model and shape optimization of an underwater glider with blended-wing-body', *International Journal of Naval Architecture and Ocean Engineering*, 2015, Vol.7, pp. 995-1006.

14. Guggilla, M., and Vijayakumar, R., 'CFD Investigation on the Hydrodynamic Characteristics of Blended Wing Unmanned Underwater Gliders with emphasis on Control Surfaces', *Proceedings of the ASME 2020 39th International Conference on Ocean, Offshore and Arctic Engineering, Volume 6B: Ocean Engineering*, Virtual, Online, August 3-7, 2020.

## 8. AUTHORS BIOGRAPHY

**Mukesh Guggilla** holds is currently a PhD Scholar at the Department of Ocean Engineering, Indian Institute of Technology Madras, India. He is a graduate of Naval Architecture and Ocean Engineering from IIT Madras and an Associate Member of the Royal Institution of Naval Architects. He is the recipient of the prestigious Prime Minister's Research Fellowship at IIT Madras. Apart from his current research area of underwater gliders, his fields of interest include Ship Design and Hydrodynamics, Biomimetic Propulsion systems and Machine Learning Design Applications for Surface vessels.

**Cdr (Dr) R. Vijayakumar** is a graduate of Naval Architecture from the Cochin University of Science and Technology. He has completed his PG Diploma in Naval Construction from Indian Institute of Technology, Delhi, and M.Tech in Ocean Engineering from Indian Institute of Technology, Madras and Ph.D. from Indian Institute of Technology, Delhi. He is presently working as Associate Professor at the Ocean Engineering Department at Indian Institute of Technology, Madras. At IIT Madras he teaches the subjects of Ship Hydrodynamics, Ship Dynamics and Advanced Marine Vehicles. He has served in the Indian Navy for period of 21 years, in dockyards and design directorate. During his tenure in the Navy, besides many other appointments, he has also served as Adjunct Faculty at the Naval Construction Wing, Department of Applied Mechanics at Indian Institute of Technology, Delhi. He has published a total of 60 technical papers in refereed Journals and Proceedings of international/National Conferences. He is a member of the Institution of Naval Architects, India. His fields of interest include Warship Design, Ship Hydrodynamics, Ship Dynamics and Computational Fluid Dynamics (CFD).

# BATTERY TECHNOLOGY – WHERE IT CAME FROM AND WHERE ITS GOING

**M Moody** and **D Mitchell**, BMT, Canada

## SUMMARY

The increasing demands placed upon the battery industry to provide battery solutions with increased power density, longer life expectancy, improved reliability, reduced maintenance and reduced through life costs has resulted in significant advances in battery chemistries. The maritime sector primarily driven by a need to reduce emissions has benefitted from these advances and has resulted in the construction of new ships being hybrid or fully electric. Numerous governments are currently in the process of renewing Naval Ships and Submarines offering an opportunity to adopt some of these technologies into the design whilst maintaining a level of assurance that safety is always priority.

## NOMENCLATURE

ABS – American Bureau of Shipping
Ah – Ampere-hours
BC – Before Christ
BMS – Battery Management System
BV – Bureau Veritas
c. – circa
DNV – Det Norske Veritas
EMS – Energy Management System
hp – Horsepower
IEC – International Electrotechnical Commission
IEEE – Institute of Electrical and Electronic Engineers
IPMS – Integrated Platform Management System
ISO – International Organization for Standards
PMS – Power Management System
LFP – Lithium Iron Phosphate
LIB – Lithium-ion Battery
LR – Lloyds Register of Shipping
LTO – Lithium Titanate
NAVSEA – Naval Sea Systems Command
NAVSO – Naval Staff Office Publication
NiZn – Nickel Zinc
NMC – Nickel Manganese Cobalt
RINA – Registro Italiano Navale
SIL – Safety Integrity Level
SwCI – Software Criticality Level
UL – Universal Laboratories

## 1. INTRODUCTION

Numerous papers have been published discussing the various battery chemistries currently available and as well as those in development. The intent of this paper is to provide a brief overview of some of the earliest historical events in the development of batteries, the electrification of ships and submarines, through to present day battery technology.

Battery technology has been a significant source of military innovation throughout recent history. It has aided changes in warfare capability from the development of portable radio communication sets to missiles. However, the driving force behind battery technology has changed due to increased demand for higher power density, extended life, and reduced maintenance to that of the more traditional lead acid cell from the automotive and renewables sectors.

This paper contributes to this discussion and the conference theme by providing an overview of battery development and considerations in naval platforms. The drive to deliver improved battery chemistries to meet the demands of the commercial sector brings opportunity for improved technologies on naval platforms. This paper investigates the installation impacts associated with integrating these new technologies in a submarine application.

## 2. A BRIEF HISTORY OF BATTERY TECHNOLOGY AND THE ELECTRIFICATION OF THE SHIPPING AND NAVAL SECTOR

The purpose of this section is to provide a brief historical timeline of significant developments in battery technology and the initial electrification of ships and submarines.

The battery is an electrochemical storage device which converts chemical energy into electrical energy. A battery is composed of several cells, connected in a manner to deliver a voltage and current for the specific requirements of the user. The chemistry used within the cell determines its cycle life and classifies the cell as either “Primary” or “Secondary”. The Primary cell relies on a chemical reaction occurring between the active materials used in its manufacture. Once this chemical reaction has been depleted it cannot be restored by recharging and the cell should be replaced. The Secondary cell chemistry requires the cell to be initially charged (commissioned) once manufactured and once depleted it can be restored by recharging (cycle).

The earliest example of a battery is dated around 250 BC with the discovery in 1938 of the Parthian or ‘Baghdad’ Battery although its purpose remains controversial. The term ‘battery’ is historically given to Benjamin Franklin who used the term in 1749 when conducting experiments using linked capacitors.

It is the Italian scientist Alessandro Volta who is credited with inventing the first battery called the voltaic pile (c.1800). This battery was based on Volta's own hypothesis of metallic electricity, which comprised of either silver and zinc (or copper and pewter) discs separated by paper or cloth moistened with either salt water (or sodium hydroxide).

In 1820 John Frederic Daniell invented the Daniell Cell, which comprised of a copper vessel filled with a copper sulphate solution and a porous pot placed inside. Inside the porous pot was a solution of dilute sulphuric acid and a zinc rod. However, throughout the 1800's, numerous chemistries and designs were being developed, but these remained to be primary batteries.

It was Gaston Planté who created the first Secondary battery in 1859 of a lead-acid type. The initial design was subsequently refined by Camille Alphonse Faure in which the original lead plates were replaced with a lead oxide paste pressed into a lead lattice.

In 1866, George Leclanché developed a battery using a Zinc Anode with a Manganese dioxide cathode and an ammonium chloride electrolyte. Leclanché later changed the liquid electrolyte to a paste resulting in the first dry cell battery. Carl Gassener in 1886, modified the chemistry by using a plaster of Paris to form an ammonium chloride paste with zinc chloride to extend the shelf life.

The electrification of shipping began in the latter part of the 1800's. This was a century of significant discovery, invention, and development for both the electrical industry and the maritime industry. This century saw the design of ships progress from the wooden hull to iron, sails to steam and in the latter part of that century to the use of the internal combustion engine. It was also a century of the electrification of ships and the advancement of submarine design forming the basis of the modern submarine.

It was not until 1880 that electricity was installed on a maritime vessel. The construction of the SS COLUMBIA is attributed as being the first commercial electrical installation onboard a maritime vessel. The vessel was installed with four steam driven Edison 'long legged Mary-Ann' dynamos one of which was used to excite the field magnets of the other three machines. The electricity was distributed via several circuits from a switchboard to 120 of Thomas Edison's '1880 wire terminal base' electric lamps.

The USS TRENTON in 1883 at Brooklyn Shipyard, was the first warship to be installed with electricity. This ship was installed with a an Armington-Sims engine, one L-Dynamo, 104 (16-candle) lamps, 130 (1-candle) lamps and 4 (32-candle) lamps [1].

This period of history saw significant advances being made in submarine design and the adoption of these new technologies. The construction of NAUTILUS in 1886, designed by Andrew Campbell and James Ash, was the first to demonstrate the benefits of an electrically driven submarine and was installed with 100 Elwell-Parker Electric Power Storage batteries. However, these required frequent recharging and overhaul [2].

In 1888, the launch of the Spanish built PERAL and the French built GYMNÔTE, heralded a new era in submarine design and use of technology.

The PERAL was launched on the 8th September 1888 and designed by Isaac Peral y Caballero, for the Spanish Navy. This submarine became one of the first electrically driven, battery powered military submarines. The propulsion system was originally to be one 40 hp electric motor driving a single propeller but was switched to two 30 hp motors driving two propellers. The initial battery design was for 430 cells but ended with 613 cells being installed of unknown chemistry [3].

Shortly after the launch of the PERAL, the GYMNÔTE was launched on the 24th September 1888. Originally developed by Henri Dupuy de Lôme but completed after his death by Gustave Zédé it was the first submarine with a military capability recognisable by today's standards in that it could launch a whitehead torpedo. This submarine was designed to be electrically driven and was installed with 564 Commelin & Desmazures alkaline cells [4] comprised of zinc and copper electrodes with a potassium hydroxide electrolyte. However, these were subsequently replaced in 1894 with 200 Laurent-Coly cells and those then replaced with 100 cells supplied by the Society of Metal working in 1897/1899 of an unknown chemistry. This submarine was one of the most successful of her era in completing over 2000 dives until running aground in 1907. However, both submarine designs were only battery powered and once depleted had to return to port for recharge.

It was the J.P. Holland submarine, Holland VI, launched in 1897 that changed the capability of the submarine and became the forerunner of the modern submarine. It was installed with a 60 cell, lead-acid battery with a 1500 Ah capacity and charged using a 45 hp Otto Gas Engine via a Dynamo which permitted the recharging of the battery without the requirement to return to port.

The first Nickel Cadmium cell was developed by Waldermar Jungner in 1899, who also invented the Nickel-Iron cell, which contained a potassium hydroxide electrolyte. This became the first alkaline cell and produced a higher energy density over the equivalent lead acid battery.

In 1901, Thomas Edison refined Jungners original Nickel-Iron cell chemistry and progressed its development using a potassium chloride electrolyte, resulting in a patented and commercially viable battery.

The diesel engine was first used on the French submarine AGIETTE in 1904. However, it wasn't until the invention of the snorkel mast in 1943 which was installed on the German submarine U-264 that allowed the operation of the submarine's engines at a shallow depth to recharge the batteries without returning to the surface.

Development of lithium battery technology began in 1912 by Gilbert Newton. However, it wasn't until the 1980's with John Goodenough discovering the Lithium Cobalt Oxide ($LiCoO_2$) cathode and Rachid Yazami discovering the graphite anode that enabled the first prototype rechargeable lithium battery to be manufactured by Asahi Chemical in Japan. However, it was not until 1991 when the first commercial lithium battery was manufactured by Sony.

## 3. SUBMARINE BATTERY APPLICATIONS

### 3.1 SUBMARINE MAIN BATTERY

The modern submarine has numerous batteries used onboard with the largest being the main battery. The main battery chemistry has been of the lead-acid type and used onboard conventional and nuclear-powered submarines. There has been little change to the lead acid battery technology since it was first used onboard submarines. However, there have been improvements in the manufacturing process, and improvements of the electrolyte chemistry, to increase capacity, extend battery endurance and overall life.

#### 3.1 (a) Diesel Electric Submarine

In a diesel electric conventional submarine application, the battery is the sole power source when submerged and supports all hotel, weapons, and propulsion loads. The propulsion load demands vary from low rates of discharge over long periods to extremely high rates of discharge for short periods to support high speed manoeuvres. The rate of discharge could then be reduced to continue submerged operations until such time as the battery is depleted.

#### 3.1 (b) Air Independent Propulsion Submarine

The Air Independent Propulsion conventional submarine is similar in size to that of the conventional diesel electric. However, it has similarities in operation akin to a Nuclear-powered submarine as it no longer requires the battery for low-speed submerged activities. However, similarly to the diesel electric conventional submarine, during high-speed submerged manoeuvres the battery is required to support high rates of discharge for short durations.

#### 3.1 (c) Nuclear-Powered Submarine

The primary source of electrical power on this type of submarine is derived from the nuclear reactor and the role of the battery changing to that of an uninterruptable power supply. In this application the battery remains in a fully charged state. In the event of a complete loss of primary power the purpose of this battery is to maintain reactor safety and provide a source of power to allow the submarine to surface.

### 3.2 ANCILLARY BATTERIES

There are various types of ancillary batteries used onboard a ship and submarine and are primarily used in an uninterruptible power supply type of application. In these applications the battery remains in a fully charged state until a loss of the primary source of supply occurs. This permits continued operation of the equipment until the restoration of normal or changeover to alternative supply. They can also support for a limited time the supplied equipment to complete operations and allow for a graceful shutdown.

## 4. BATTERY TECHNOLOGIES

In recent years, significant advances in battery technology have occurred to meet the demands primarily from the automotive, power generation and renewable energy sectors. These demands have necessitated improvements to performance (power density), extended life expectancy, reducing maintenance, weight and overall through-life-costs compared to that of the lead acid battery and in a submarine application reduce the indiscretion ratio.

The efforts by the battery manufacturers to meet these demands has seen the emergence and refinement of numerous electrical energy storage solution technologies. These technologies are continuing to be developed to improve performance, reliability, increase safety and reduce costs, making these more viable alternatives to traditional lead acid batteries.

The continuing development and improvements in Lithium-ion battery (LIB) technology, is proving a viable alternative to that of lead acid, depending on the application. However, LIB technology comprises of several chemistries, which are continuously being refined to further improve power density, reliability, safety and extending the life of the product.

### 4.1 SUBMARINE APPLICATION

Within the naval submarine environment, lead acid battery technology remains the favoured electrical energy storage solution due primarily to safety concerns surrounding LIB technology and thermal runaway. However, various navies and submarine manufacturers are actively investigating alternative technologies as a replacement to the traditional lead acid battery. Some of the technologies currently being investigated as alternatives to LIB technology is Nickel Zinc and Silver Zinc chemistries. Table 1 provides a selection of navies and submarine manufacturers presently pursuing or adopting alternative battery technologies for submarine main battery applications.

**Table 1 – Navies using alternative technologies**

| Nation | Manufacturer | Chemistry | Submarine Class | Notes |
|---|---|---|---|---|
| Japan | GS Yuasa | Lithium nickel cobalt aluminium oxide | Soryu Class Batch II | Chosen as preferred supplier to Toshiba during bidding phase |
| | Toshiba | Lithium-titanite | | Considered for Soryu Class |
| Republic of Korea | Samsung SDI | Lithium chemistry undisclosed | KSS-III | In build |
| Germany | ThyssenKrupp/Saft | Lithium Super Phosphate | HDW Class 212/214 | Possible chemistry Lithium iron phosphate |
| France | Naval Group | Lithium chemistry undisclosed | | Assumed Lithium technology |
| USA/UK | Enersys | Nickel Zinc | | Research |
| Italy | FAAM | Lithium Phosphate | U212 NFS | Future submarine |
| China | Unknown | Unknown | Yuan Class | Reports suggest lithium battery technology |
| Australia | Naval Group | Lithium chemistry undisclosed | | Potential for Collins Class Replacement future submarine |
| Sweden | SAAB Kockums | Lithium chemistry undisclosed | | Research into a replacement and future submarine battery |
| Russia | Unknown | Unknown | Loshark | Reported as lithium battery technology |

## 4.2 COMMERCIAL INDUSTRY

In a drive to reduce emissions and having similar demands as the automotive and renewable energy sector industry of increased power density, reduced maintenance and extended life. There has been significant investment within the maritime sector in the development of hybrid, fully electric and zero emission vessels. Table 2 provides a small selection of manufactures providing battery solutions to the commercial marine sector or used in other military applications.

**Table 2 – Alternative Battery Technology Suppliers**

| Manufacturer | Chemistry | Additional information |
|---|---|---|
| Kokam | NMC+LFP+LTO (NANO) | Specially designed for defense & aerospace application<br><br>This hybrid type cell has incorporated the advantages of NMC, LFP and LTO cells in one cell. |
| Epsilor | Lithium-ion, Lithium Polymer, Nickel Metal Hydride and Nickel Zinc | Military Vehicles, Dismounted Soldier, UAVs, Submarines |
| Tadiran | Lithium Thionyl Chloride, Lithium Metal Oxide and Lithium-Ion, Hybrid Layer Capacitor | Military Vehicles, Naval, UAVs, Dismounted Soldier |
| Denchi Power | 6T Lithium-ion | Military Vehicles, UAVs, Dismounted Soldier |
| Revision Military | Lithium-ion | Military Vehicles, Dismounted Soldier. Manufacturing partner Kokam |
| UltraLife | Lithium-ion, Lithium Polymer | Dismounted Soldier, Military Vehicles, UAVs |
| RS Aqua | Lithium-ion | Commercial Submersibles, Commercial Subsea applications |
| General Atomics | Lithium-ion Fault Tolerant (LIFT), Aluminium Power System (ALPS) | Research and Development for Land, Air, Manned Submersibles, Unmanned Submersibles<br>LIFT - ALPS – fuel cell technology |
| SubCTech | Lithium-ion | Remote Operated Vehicle, Autonomous Underwater Vehicle, Commercial Subsea applications, |
| Siemens/FREYR | Lithium Chemistry Undisclosed | Renewables, Marine and Electric Vehicle |
| Corvus Energy/Siemens | Lithium Chemistry Undisclosed | Commercial Shipping and offshore industry (Energy Storage System (ESS)) |
| Sterling PlanB | NMC | Commercial Shipping, battery module completed A60 IMO FTP code without thermal runaway |
| Ficantieri SI | Lithium Chemistry Undisclosed | Commercial Shipping (ESS on two cruise ships) |
| Kongsberg (originally developed by Rolls Royce SAVe Energy hybrid system) | Lithium Chemistry Undisclosed | Commercial Shipping (Electric Hybrid System) |
| ZAF | NiZn | Group 31 batteries undergoing testing for military application by USN |

The advances in LIB chemistries and manufacturing processes has resulted in the increase of power density, improved safety and overall reduced through life costs associated with this technology and as such making it the preferred solution across most sectors. In a maritime and more specifically a submarine domain, it is still relatively new and there are limited numbers of vessels adopting this technology due to the potential risk of failure and management of a thermal runaway event.

## 4.3 COMPARISION OF TECHNOLOGIES

An example comparing three battery chemistries, lead-acid, lithium-ion and nickel zinc is shown in Figure 1 [5]. This figure does not differentiate between the various chemistries of lithium-ion, to which these will be shown in subsequent figures. However, it does offer a comparison of how each of these technologies performs against those criteria associated with a battery. The further from the centre of the radar diagram shows improved performance in meeting these criteria.

**Figure 1 – Battery Performance Comparison**

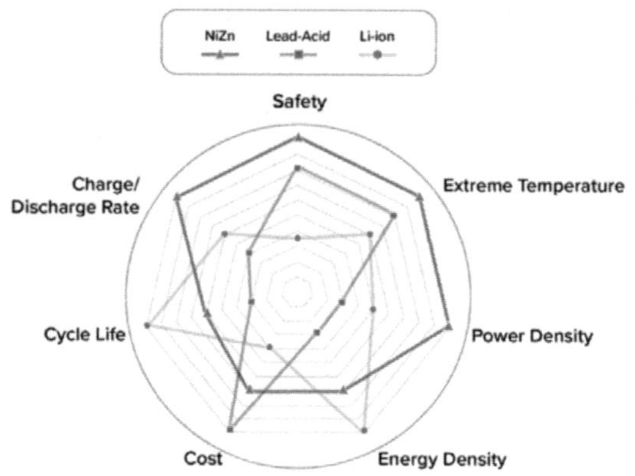


## 4.4 COMPARISON OF LITHIUM CHEMISTRIES

There are various lithium-ion chemistries which are continually being refined or new chemistries being developed, as are the manufacturing processes. The specifics relating to these chemistries and the manufacturing processes are proprietary information. The illustrations shown in Table 3 [6] [7] are provided to give an example of how each of these chemistries perform based on the criteria below and provide examples of the applications in where generally used:

- Specific Energy/Energy;
- Specific Power/Power;
- Safety;
- Performance;
- Life Span; and
- Cost.

Table 3 – Comparison of Lithium-ion Chemistries

| Chemistry | Chemical Symbols | Common Applications | Performance Characteristics |
|---|---|---|---|
| Lithium Cobalt Oxide (LCO) | $LiCoO_2$ | Mobile phones, laptops and digital cameras | 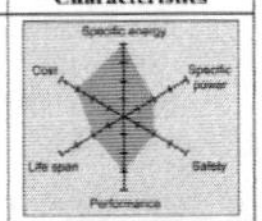 |
| Lithium Manganese Oxide (LMO) | $LiMn_2O_4$ | Power tools, medical instruments, hybrid and electric vehicles | 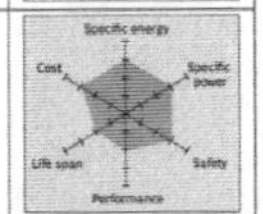 |
| Lithium Nickel Manganese Cobalt Oxide (NMC) | $LiMnCoO_2$ | Electric vehicles | 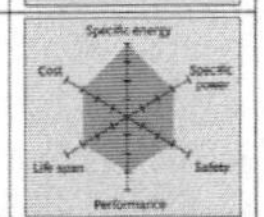 |
| Lithium Iron Phosphate (LFP) | $LiFePO_4$ | Vehicle lead acid starting battery replacement. | 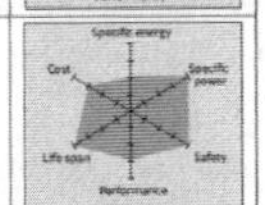 |
| Lithium Nickel Cobalt Aluminium Oxide (NCA) | $LiNiCoAlO_2$ | Hybrid and electric vehicles | 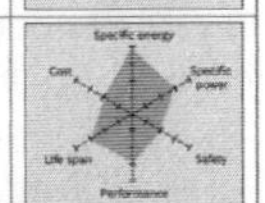 |
| Lithium Titanate (LTO) | $Li_2TiO_3$ | Electric vehicles, UPS and Solar-powered street lighting | 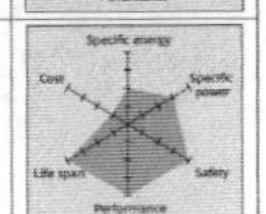 |
| Kokam proprietary chemistry comprising a hybrid type cell comprising:<br><br>Lithium Nickel Manganese Cobalt Oxide + Lithium Iron Phosphate + Lithium Titanate | NMC+LFP+LTO (NANO) | Specially designed for defense & aerospace application Suitable for extremely volatile and dynamic operational conditions. The high power, energy and safety features allow the NANO cells to be flexibly applied in various applications. | 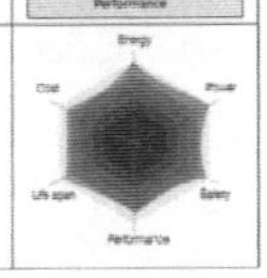 |

There are four chemistries identified in Table 3 that stand out as offering the most favourable performance characteristics, these being Lithium Nickel Manganese Cobalt Oxide (NMC), Lithium Iron Phosphate (LFP), Lithium Titanate (LTO) and the NMC+LFP+LTO (NANO) hybrid cell.

Comparing these four chemistries performance characteristics to those known chemistries identified in Table 1 as being used by Navies, LFP and LTO are the favoured choice of chemistry for submarine main battery solutions at this time.

## 5. STANDARDS

Several Navies have, or are in the process of, adopting standards of the commercial industry and various classification societies rules relating to naval ship and submarine designs, with additional military requirements being retained within the military standards.

The increased use of primarily lithium-ion battery technology onboard commercial ships has necessitated the classification societies to produce additional rules and guidance documentation relating specifically to these types of installations. Table 4 provides a list of those rules specifically for naval ship, submarine and lithium-ion batteries.

Table 4 – Classification Society Rules

| Classification Society | Rules |
|---|---|
| ABS | Guide for the Building and Classing of International Naval Ships |
| | Guide for use of lithium batteries in the marine and offshore industry |
| BV | NR 483 Rules for the Classification of Naval Ships |
| | BV NR535 Rules for Classification of Naval Submarines |
| DNV | DNVGL-RU-NAVAL Rules for the Classification of Naval vessels, Part 3 Surface Ships |
| | DNVGL-RU-NAVAL, Part 4 Sub-surface ships, Chapter 1 Submarines |
| | DNVGL-CP-0418 Type Approval Lithium-ion batteries |
| | Technical Reference for Li-ion Battery Explosion Risk and Fire Suppression |
| LR | Rules and Regulations for the Classification of Naval Ships |
| | Test Specification 5 Type testing for lithium battery systems |
| RINA | Rules for the Classification of Naval Ships |
| | NCC92-ENG Rules for the Certification, Installation and Testing of Lithium Based Storage Batteries |

The classification society published documents identified in Table 4 for Naval Ships provide similar guidance battery installations and where commonality exists reference to commercial shipping rules. The BV and DNV Rules for submarines also have similar guidance for battery installations and where commonality exists refer to their respective Naval Ship Rules. The subject areas contained within these rules can be identified as being:

- Quality Assurance
- Safety
- Reliability
- Battery Management Systems (BMS)
- Energy Management System (EMS) or Power Management System (PMS)
- Installation Requirements
- Protection

The classification society guidance documents and rules for lithium batteries refer to specific test qualification requirements of the standards identified in Table 5.

Table 5 – Commercial Standards

| Standard | Title | Classifications Society |
|---|---|---|
| IEC 62619 | Secondary cells and batteries containing alkaline or other non-acid electrolytes – Safety requirements for secondary lithium cells and batteries for use in industrial applications | ABS, DNVGL, LR, RINA |
| IEC 62620 | Secondary cells and batteries containing alkaline or other non-acid electrolytes – secondary lithium cells and batteries for use in industrial applications | ABS, DNVGL, LR, RINA |
| IEC 60079-10-1 | Explosive atmospheres Part 10-1: Classification of areas – Explosive gas atmospheres | ABS |
| IEC 60092-101 | Electrical installations in ships | DNVGL |
| IEC 62281 | Safety of primary and secondary lithium cells and batteries during transportation | ABS |
| IEC 62660 | Secondary lithium-ion cells for the propulsion of electric road vehicles – Part 1: Performance Testing | ABS |
| ISO 12405-3 | Electrically propelled road vehicles – Test specification for lithium-ion traction battery packs and systems – Part 3: Safety performance requirements | |
| IEC 62881 | Cause and effect matrix | ABS |
| UL 1642 | Standard for lithium batteries | ABS |
| UL 2054 | Standard for household batteries | ABS |
| SOLAS | International Convention for the Safety of Life at Sea | ABS |
| International Maritime Organization (IMO) | International Maritime Dangerous Goods (IMDG), Dangerous Goods List, Substance Details – UN 3481, Lithium-ion Batteries Contained in Equipment (including lithium-ion polymer batteries) | ABS |
| USCG 46 CFR Subchapter J, Electrical Engineering | United States Coast Guard | ABS |
| IEEE 45 (series) | IEEE Recommended practice for electrical installations on shipboard | |
| IEEE 1679 | IEEE Guide for the characterization and evaluation of lithium-based batteries in stationary applications | |

In addition to these standards there are some military specific standards relating to batteries produced by the US Military as shown in Table 6 to which the classification society ABS refer to in their documentation.

Table 6 – US Military Standards

| Number | Title | Classifications Society Referenced |
|---|---|---|
| NAVSEA TM-S9310-AW-SAF-010 | US Navy Technical Manual for Batteries, Navy Lithium Safety Program Responsibilities and Procedures | ABS |
| NAVSEA SG270-BV-SAF-010 | High-Energy Storage System Safety Manual | ABS |
| NAVSO P-3676 | Best Manufacturing Practices Navy Primary and Secondary Batteries | |

As can be seen from the above Tables there are several different standards referred to by the classification societies with the most commonly referenced standards being IEC 62619 and IEC 62620 for cell and module testing.

## 6. SUBMARINE INSTALLATIONS

### 6.1 QUALITY ASSURANCE

The classification society and commercial standards relating to all batteries and onboard installations place significant emphasis in the ability to demonstrate a high level of assurance. This requires the implementation of a robust quality management system and testing regime to demonstrate that the battery is being manufactured consistently to a high quality with minimal potential of defects. It also encompasses the physical installation onboard the submarine and the ability of the onboard ancillary systems necessary to support the battery during normal operations or in the event of a fault occurring.

To demonstrate this level of assurance and based on some of the high-level documented evidence required by the classification societies directly relating to the battery system include:

- Circuit and block diagrams of all battery system components;
- General arrangement drawings;
- Detailed system functional description and operational limitations;
- Software quality requirements;
- Gas analysis testing;
- Identity of the test establishment and accreditation;
- Block diagram detailing the equipment under test configuration;
- List of tests demonstrating compliance with the relevant tests outlined and specified standards; and
- Certificates and reports relevant to tests previously obtained for the product.

## 6.2 BATTERY TESTING REQUIREMENTS

The level of testing required to be conducted by the classification societies necessary to support the installation of a battery and if choosing a lithium-ion battery solution are shown in Table 7 [8].

Table 7 – Cell and Battery Testing

| Item | Cell | Battery Module / System |
|---|---|---|
| External short circuit test (IEC 62619 – Section 7.2.1) | X | X |
| Impact Test (IEC 62619 -Section 7.2.2) | X | |
| Drop Test (IEC 62619 – Section 7.2.3) | X | |
| Thermal abuse test (IEC 62619 – Section 7.2.4) | X | |
| Overcharge test (IEC 62619 – Section 7.2.5) | X | |
| Forced discharge test (IEC 62619 – Section 7.2.6) | X | |
| Internal short circuit test (IEC 62619 – Section 7.2.3) (optional) | X | |
| Propagation test (IEC 62619 – Section 7.3.3) | | X |
| Overcharge control test (IEC 62619 – Section 8.2.2 & 8.2.3) | | X |
| Overheating control test (IEC 62619 – Section 8.2.4) | | X |
| Charging procedure for test purposes (IEC 62620 Section 6.2) | | X |
| Discharge performance at: (IEC 62620 – Section 6.3.1) | X | |
| Discharge at low temperature (IEC 62620 – Section 6.3.2) | X | |
| High rate permissible current (IEC 62620 – 6.3.3) | X | |
| Capacity retention and recovery (IEC 62620 – Section 6.4) | X | |
| Internal resistance test (IEC 62620 – Section 6.5) | X | X |
| Endurance Tests (IEC 62620 – Section 6.6) | | X |
| Gas Analysis | X | |
| Cell balancing functional test | | X |
| Sensor failures test | | X |
| SoC validation test | X | X |
| SoH validation test | X | X |
| High voltage test | | X |
| Insulation resistance test | | X |
| Emergency disconnect functional test | | X |
| Alarms and shutdowns test | | X |
| Over-temperature protection test | | X |
| Over-voltage protection test | | X |
| Under-voltage protection test | | X |
| Communication failure between BMS and charging system shutdown test | | X |

Although the tests identified in the table above were specifically developed for lithium-ion batteries, certain aspects still apply to other battery chemistries.

## 6.3 FAILURE MODES AND EFFECTS ANALYSIS

Due to the higher energy density of alternative battery technologies compared to lead-acid, a failure modes and effects analysis of the battery, the battery systems and the ancillary systems used to support the battery onboard the submarine should be conducted. This would apply to original design or retrofit of an alternative battery chemistry. It would identify the hazards associated for a cell, battery or battery system failure and the impact this may directly have on the electrical power system or any ancillary system, identify mitigation measures that could be taken to ensure the safety of the submarine and its personnel.

The submarine areas and systems which should be considered as being impacted are:

### 6.3 (a) Power Generation and Conversion System

The battery technology chosen would have a different charge and discharge profile to that of the traditional lead acid battery. The level of monitoring and control to maintain the optimal operating parameters of the battery would need to be assessed. If retrofitting to an existing platform, consideration should be given to the capability of the generation and power conversion systems in supporting an increased charging current if the battery can support rapid charging.

### 6.3 (b) Control Systems

The integration of control systems used for power generation, power conversion, Power Management System or Integrated Platform Management System should be assessed. This would be to determine the level of monitoring and control requirements to ensure safe operating parameters are maintained within operational limits or in the event of a battery fault reduce or stop the charge or discharge and if necessary, initiate the protection devices.

For ancillary battery applications the capabilities of the existing charging circuit and its associated control system integration would need to be assessed.

### 6.3 (c) Generator Prime Movers

The capability of the legacy generator prime movers (e.g., diesel engine) and its capability to support the potential increase in load demand resulting from the increase in electrical load demand placed on the generator would require assessment.

### 6.3 (d) Power Distribution System

The higher energy density and increase in charging current requires assessment especially when considering retrofit into a legacy submarine design. Assessment should be made of the current carrying capacity of the system from generators to battery and battery to battery switchboard. An additional consideration would be to the distribution systems protection devices and their capability to limit or clear battery fault currents.

Integration of the Battery Management System with those protective devices installed immediately at the battery side of the switchboard should be a consideration to potentially increase the safety margin and isolate the battery prior to a failure occurring.

### 6.3 (e) Atmosphere Monitoring and Smoke Detection

The capabilities of the onboard atmosphere monitoring and smoke detection systems require assessment to confirm their ability to detect the range of gases produced, the number installed and locations within the battery compartment.

The gases produced by a failed cell/battery differ to those produced by a lead-acid battery.

6.3 (f) Fire Suppression

The method of fire suppression within the battery compartment should be evaluated as to its suitability in extinguishing and inhibiting the propagation of a catastrophic cell failure of the type of battery chemistry being installed.

6.3 (g) Battery Cooling/Battery Ventilation

The battery cooling water system and its equipment (if fitted) would require assessment to verify if the cooling system could provide sufficient cooling water flow and identify what level of adaptation would be required to the system in order to accommodate and maintain the optimum operating temperature of the battery modules.

The battery ventilation system and its equipment will require assessment in providing the optimum cooling airflow to maintain the correct operating temperature of the battery. The cooling airflow for a lead-acid battery compartment dissipates the heat and gases generated by moving the air over the tops of the cells. However, for other battery technologies and layout within the compartment may require a different airflow path to ensure the required cooling.

The assessment should also consider if integration of the ventilation control system or IPMS with the BMS. This could permit the BMS to automatically isolate the compartment and the compartment atmosphere in the event of a cell/module failure.

The capability of ventilation ducting and isolation valves to contain gases produced or potential shockwave as a result of a catastrophic failure should also be assessed.

The ventilation system design should also be assessed to ensure the safe containment of the gases produced by a cell or battery failure when ventilating the compartment post cell/module failure.

6.3 (h) Compartment Design

The lead-acid battery is extremely heavy and provides ballast weight to the submarine. This has meant that the battery has been positioned low within the submarine and around the centre line, with the cells laid out on the floor and on multiple tiers as determined by the hull form.

Alternative cell technologies can be significantly lighter and no longer required to be placed side by side like a lead-acid battery. These could be rack mounted and stacked vertically allowing access for maintenance between the racks.

On a legacy platform this may cause stability issues and require additional ballasting arrangements. Whereas on a new design platform the position of the battery compartment could be elsewhere on the submarine and allow for larger or heavier equipment or tanks to be positioned in that region of the submarine.

Depending on the battery chemistry chosen an assessment of the compartments suitability to withstand and contain a catastrophic cell failure should be considered.

6.3 (i) Electromagnetic Compatibility

Electromagnetic Compatibility testing should be carried out to ensure safe, reliable operation and monitoring of the battery when installed on the submarine. An example of the testing requirements is provided in Table 8 [9].

Table 8 – Electromagnetic Compatibility Tests

| Equipment and Subsystems Installed In, On, or Launched From the following Platform or Installation | Requirement Applicability | | | | | | | | | | | | | | | | | | |
|---|---|---|---|---|---|---|---|---|---|---|---|---|---|---|---|---|---|---|---|
| | CE101 | CE102 | CE106 | CS101 | CS103 | CS104 | CS105 | CS109 | CS114 | CS115 | CS116 | CS117 | CS118 | RE101 | RE102 | RE103 | RS101 | RS103 | RS105 |
| Submarine | A | A | L | A | S | L | S | L | A | S | L | S | S | A | A | L | L | A | L |
| Legend<br>A - Applicable<br>L – Limited as specified in the individual sections of this standard<br>S – Procuring activity must specify in procurement documentation | | | | | | | | | | | | | | | | | | | |

The tests identified in the Requirements Matrix are titled as follows:

CE101, conducted emissions, audio frequency currents, power leads
CE102, conducted emissions, radio frequency potential, power leads
CE106, conducted emissions, antenna port
CS101, conducted susceptibility, power leads
CS103, conducted susceptibility, antenna port, intermodulation
CS104, conducted susceptibility, antenna port, rejection of undesired signals
CS105, conducted susceptibility, antenna port, cross modulation
CS109, conducted susceptibility, structure current
CS114, conducted susceptibility, bulk cable injection
CS115, conducted susceptibility, bulk cable injection, impulse excitation
CS116, conducted susceptibility, damped sinusoidal transients, cables and power leads
CS117, conducted susceptibility, lightning induced transients, cables and power leads
CS118, personnel borne electrostatic discharge
RE101, radiated emissions, magnetic field
RE102, radiated emissions, electric field
RE103, radiated emissions, antenna spurious and harmonic outputs
RS101, radiated susceptibility, magnetic field
RS103, radiated susceptibility, electric field
RS105, radiated susceptibility, transient electromagnetic field

The purpose of this testing is to establish and control the electromagnetic emissions and determine the susceptibility of the electrical, electromechanical equipment and sub systems being designed for use in a submarine. This is particularly relevant in ensuring the safety of the LIB and that the BMS can continue operating during submarine deperming activities, where high level magnetic fields are present.

The submarine power system is ungrounded and the capacitance to ground of power filters provides a path for conducting current into the hull. These currents can result in magnetic fields being generated which can affect other sensitive equipment within the hull of the submarine and impact the submarines magnetic signature. In addition, these tests are also used to limit the total line to ground capacitance of the ungrounded electrical system to maintain the performance of the onboard ground fault detection system and limit the hull currents and line to ground voltages.

6.3 (j) Safety Integrity Level

The Safety Integrity Level (SIL) of the BMS, IPMS and PMS requires consideration.

This is to ensure the BMS, IPMS and PMS have a demonstrable level of integrity and reliability in maintaining and supporting the battery and those ancillary submarine systems necessary for the safe operation of the battery installed onboard a submarine.

Applying the guidance contained within the US MIL-STD-882 System Safety. the Software Control Category of the BMS for a main or ancillary battery could be considered as level 1 as defined in Table 9.

Table 9 – Software Control Category

| Level | Name | Description |
|---|---|---|
| 1 | Autonomous (AT) | Software functionality that exercises autonomous control authority over potentially safety-significant hardware systems, subsystems, or components without the possibility of predetermined safe detection and intervention by a control entity to preclude the occurrence of a mishap or hazard.<br>*(This definition includes complex system/software functionality with multiple subsystems, interacting parallel processors, multiple interfaces, and safety-critical functions that are time critical.)* |
| 2 | Semi-Autonomous (SAT) | Software functionality that exercises control authority over potentially safety-significant hardware systems, subsystems, or components, allowing time for predetermined safe detection and intervention by independent safety mechanisms to mitigate or control the mishap or hazard.<br>*(This definition includes the control of moderately complex system/software functionality, no parallel processing, or few interfaces, but other safety systems/mechanisms can partially mitigate. System and software fault detection and annunciation notifies the control entity of the need for required safety actions.)*<br><br>Software item that displays safety-significant information requiring immediate operator entity to execute a predetermined action for mitigation or control over a mishap or hazard. Software exception, failure, fault, or delay will allow, or fail to prevent, mishap occurrence.<br><br>*(This definition assumes that the safety-critical display information may be time-critical, but the time available does not exceed the time required for adequate control entity response and hazard control.)* |
| 3 | Redundant Fault Tolerant (RFT) | Software functionality that issues commands over safety-significant hardware systems, subsystems, or components requiring a control entity to complete the command function. The system detection and functional reaction includes redundant, independent fault tolerant mechanisms for each defined hazardous condition.<br>*(This definition assumes that there is adequate fault detection, annunciation, tolerance, and system recovery to prevent the hazard occurrence if software fails, malfunctions, or degrades. There are redundant sources of safety-significant information, and mitigating functionality can respond within any time-critical period.)*<br><br>Software that generates information of a safety-critical nature used to make critical decisions. The system includes several redundant, independent fault tolerant mechanisms for each hazardous condition, detection and display. |
| 4 | Influential | Software Generates information of a safety-related nature used to make decisions by the operator, but does not require operator action to avoid a mishap. |
| 5 | No Safety Impact | Software functionality that does not possess command or control authority over safety-significant hardware systems, subsystems, or components and does not provide safety-significant information. Software does not provide safety-significant or time sensitive data or information that requires control entity interaction. Software does not transport or resolve communication of safety-significant or time sensitive data. |

Using the Software Control Category of Level 1 and using the definition for Software safety criticality shown in Table 10, could be defined as SwCI 1 under the Catastrophic Severity Category.

Table 10 – Software Safety Criticality Matrix

| Software Control Category | Severity Category | | | |
|---|---|---|---|---|
| | Catastrophic (1) | Critical (2) | Marginal (3) | Negligible (4) |
| 1 | SwCI 1 | SwCI 1 | SwCI 3 | SwCI 4 |
| 2 | SwCI 1 | SwCI 2 | SwCI 3 | SwCI 4 |
| 3 | SwCI 2 | SwCI 3 | SwCI 4 | SwCI 4 |
| 4 | SwCI 3 | SwCI 4 | SwCI 4 | SwCI 4 |
| 5 | SwCI 5 | SwCI 5 | SwCI 5 | SwCI 5 |

| SwCI | Level of Rigour Tasks |
|---|---|
| SwCI 1 | Program shall perform analysis of requirements, architecture, design, and code; and conduct in-depth safety-specific testing. |
| SwCI 2 | Program shall perform analysis of requirements, architecture, and design; and conduct in-depth safety-specific testing. |
| SwCI 3 | Program shall perform analysis of requirements and architecture; and conduct in-depth safety-specific testing. |
| SwCI 4 | Program shall conduct safety-specific testing. |
| SwCI 5 | Once assessed by safety engineering as Not Safety, then no safety specific analysis or verification is required. |

However, a full and formal analysis should be conducted to accurately confirm the level of safety criticality, determine the Safety Integrity Level (SIL) associated with the BMS, IPMS and PMS and the associated onboard support systems necessary to safely operate and maintain the chosen battery chemistry.

6.3 (k) Legacy Submarine Installations

The replacement of a lead-acid battery on a legacy submarine for an alternative battery chemistry, may result in the following systems no longer being required:

- Hydrogen clearing fans;
- Distilled water (topping up) system and tank;
- Battery agitation system;
- Hydrogen detection system;
- Hydrogen eliminators; and
- Hydrogen burners (if fitted).

The battery exhaust system (if fitted) may also be considered redundant, but if after assessment may be considered suitable for compartment atmosphere clearing in the event of a battery failure.

Additional systems may also be affected, such as battery cooling, but this would need to be assessed on a system-by-system basis.

6.4 ADDITIONAL SAFETY SUPPORT SYSTEMS

Depending on the battery chemistry chosen for installation on a submarine would determine the level of additional systems required to maintain the safe operation of that battery technology. Consideration for these additional systems, has been based on classification societies guidance.

6.4 (a) Battery Management System

The Battery Management System (BMS), should provide the following control and monitoring parameters:

Control;

- Provide limits for charging and discharging to the battery converter;
- Protect against over-current, over-voltage and under-voltage by disconnection of the battery system;
- Protect against over-temperature by disconnection of the battery system; and
- Provide cell and module balancing.

Monitoring;

- Provide warning and alarm notification of cell voltage, temperature and current in a manned location.

These parameters shall be indicated at local control panels or on remote panels.

6.4 (b) Energy Management System

The Energy Management System (EMS) may be a separate system or be combined within the BMS. The function of the EMS is to monitor the voltage, current and temperature of the cells/battery during discharge and recharge cycles.

The EMS monitors these parameters to ensure the cells/battery is maintained within design parameters and reduce risk of failure due to over-discharging, excessive rate of discharge, excessive recharge rate and overcharging.

The EMS could control directly or be integrated with the Power Management System (PMS) or Integrated Platform Management System (IPMS) to maintain the output of the charging system and manage the load demands of the distribution system. The full integration requirements for this will require further assessment.

6.4 (c) Damage Control

The damage control documentation is developed from the submarines Concept of Operations (CONOPS) by the Naval Authority and determines the submarines tasks, abilities, including foreseeable damage survivability, maintenance philosophy and environmental conditions.

The CONOPS documentation defines the electrical requirements to maintain the safety of the submarine and the level of system redundancy, diversity and capacity to ensure propulsion and manoeuvrability during normal operations and during a failure condition.

The CONOPS requirements in part, are supported in the development of the FMEA completed for each of the submarine systems. The FMEA conducted provides some of the mitigating actions in the event of equipment or system failures.

## 7. CONCLUSIONS

The alternative battery technologies being adopted into the modern submarine design are not necessarily new or novel as most were discovered over a hundred years ago.

It has been due to the more recent advances in cell chemistries and manufacturing technology, driven by increased demand for higher power density, extended life and reduced maintenance that these technologies have evolved into being more viable alternatives to the traditional lead-acid batteries.

However, careful consideration and assessment of these technologies should be conducted to ensure the suitability of these in a submarine environment. This is due these technologies offering significantly higher energy densities and bring with them other hazards and risks to that of the lead-acid battery. The knowledge and understanding of the risks and hazards when humans are exposure to the gases released in the event of a failure as well as the cell chemistries' fault tolerance need to be fully assessed.

These alternative battery technologies may produce higher short circuit levels and as such the ability of the protection devices to carry and clear these faults must be fully understood.

The integration of these alternative battery chemistries in a naval submarine is not yet standardized to fully address these hazards.

For legacy submarine designs retrofitting an alternative battery technology presents additional risks, hazards and costs when the originally design was for lead-acid battery. Consideration should also be given to the various systems impacted by the change to an alternative battery technology. These include Weight and Stability, Power Distribution, Cooling, IPMS / PMS Interface,

Atmosphere Monitoring, EMC and renders some systems no longer necessary and therefore redundant.

Consequently, changing from lead acid to an alternative battery chemistry on a legacy submarine may increase the level of risk in terms of safety, technical, operation and program aspects.

However, giving careful consideration and thorough assessment alternative battery chemistry can be incorporated into various aspects of legacy and new design submarines.

As recently demonstrated by Japan and the Republic of Korea navies the integration of lithium-ion battery technology onboard a naval submarine can be achieved.

## 8. REFERENCES


1. Naval History and Heritage Command https://www.history.navy.mil/

2. Wolverhampton History & Heritage Website http://www.historywebsite.co.uk

3. Historia y Argueologia Maritima http://www.histarmar.com.ar/InfGral/AASidoli/SubmarinoPeral.htm

4. Internet Archive https://archive.org/details/jstor-1764764/page/n1/mode/2up, Science Vol. XIII No. 320, March 22, 1889

5. ZincFive https://www.zincfive.com

6. Battery University https://batteryuniversity.com/

7. Kokams http://kokam.com/

8. LR Test Specification 5 – Type testing for lithium battery systems https://www.lr.org/en/type-approval-test-specifications/

9. MIL-STD-461G – Requirements for the control of Electromagnetic Interference Characteristics of Subsystems and Equipment


## 9. AUTHORS BIOGRAPHY

**Martin Moody** is an Electrical Engineer with 17 years experience in marine electrical engineering in Australia, Canada and the UK. He holds master's and bachelor's degrees in electrical engineering. Martin has completed whole vessel electrical design for hovercrafts and rescue submarines for the Swedish Navy and Coast Guard, as well as major upgrades, analyses and sustainment projects for ships and submarines of the Australian and Canadian Navies, as well as Coast Guards and ferry operators.

**Derrick Mitchell** is a Senior Electrical Technologist with over 25 years' experience within naval marine electrical engineering, specializing in submarines. Derrick career began as an electrical fitter supporting the UK Royal Navy submarines at Rosyth Royal Dockyard. He then took an appointment with the UK Ministry of Defence as the Sub-Section Head with Delegated Design Authority responsibility supporting the Nuclear Submarine Fleet Power Generation, Distribution and Storage systems for Swiftsure, Trafalgar, Vanguard, Astute and initial design for Successor Project (Dreadnought) Class submarines. Most recently, his career has seen him move to Canada where as a contractor to the Department of National Defence he provides support to the Victoria Class Submarine Project on all electrical design aspects.

**The Royal Institution of Naval Architects**
8-9 Northumberland Street
London, WC2N 5DA
United Kingdom

ISBN 978-1-7138-8308-1

CURRAN ASSOCIATES INC.
proceedings.com